ENVIRONMENTAL RESEARCH ADVANCES

BIOLOGICAL INVASIONS

PATTERNS, MANAGEMENT AND ECONOMIC IMPACTS

ENVIRONMENTAL RESEARCH ADVANCES

Additional books in this series can be found on Nova's website under the Series tab.

Additional e-books in this series can be found on Nova's website under the e-book tab.

ENVIRONMENTAL RESEARCH ADVANCES

BIOLOGICAL INVASIONS

PATTERNS, MANAGEMENT AND ECONOMIC IMPACTS

REBECCA WATERMAN
EDITOR

nova publishers
New York

Library of Congress Cataloging-in-Publication Data

ISBN: 978-1-63482-019-6

Published by Nova Science Publishers, Inc. † New York

CONTENTS

PREFACE

Alien—also called non-indigenous or non-native—species are defined as those species that colonize an area beyond their natural range, where they reproduce and establish a population. It is known that plants, animals and microorganisms have been intentionally transferred with human discovery voyages for centuries and, possibly, since the shift from foraging to agricultural and pastoralist societies. As a result, many 'exotic' species are now among our preferred foods, dearest pets, good-looking houseplants and decorative aquarium weeds, but most of them cannot be considered alien species, as they grow well only in artificially-controlled conditions or in protected environments. Conversely, alien species are capable of ready acclimation to novel habitats, where they may find themselves unconstrained by the limiting factors—both abiotic and biotic—typical of their original habitat. In these 'favorable' conditions, they might outgrow, and ultimately overthrow, resident organisms with which they happen to compete: in this case, alien species are often dubbed 'invasive'. This book discusses patterns, management and economic impact of these biological invasions.

Chapter 1 – Alien species, defined as those species that are introduced to an area outside their natural range, are recognized as a major threat to the conservation of biodiversity and functioning of marine ecosystems. Coastal marine habitats are among the most heavily invaded systems on earth, which is derived, in part, by human-mediated transport of potentially invasive species in ships' ballast as well as intentional introductions for aquaculture and fisheries enhancement. Transfers of cultured fish and shellfish are also responsible for the accidental introduction of 'hitch-hiking' species, which could become invasive and severely affect the recipient ecosystem. In addition, man-induced climate change and eutrophication are expected to increase the susceptibility of ecosystems to invasion by alien species.

Although microorganisms are by far the most abundant and diverse of aquatic life forms, research on microbial alien species introductions has lagged behind that on metazoans and macroalgae. It is now evident, however, that due to human activities, thousands of species of marine viruses, prokaryotes and protists are being transported within and between oceans, irrespective of any natural barriers. Because of their metabolic plasticity, short generation times, wide tolerance to environmental conditions, and ability to form resting stages, microorganisms are likely the most successful invaders of coastal ecosystems.

It is well known that invasive microalgae capable of forming massive blooms can cause economic losses to aquaculture, fisheries and tourism activities and have important environmental and human health impacts. There is also growing evidence that, besides free-living species, pathogens and parasites are being frequently introduced into coastal systems, with potentially enormous consequences for economic activities and human health. Regulations at national and above-national levels are thus being implemented to face these emergent issues.

Chapter 2 – In last centuries, many bird species have been introduced worldwide. Some of these bird species have become successful invaders in the new areas, thus representing a serious problem for wildlife, agriculture and even human health. Parasites have been proposed to play a major role on bird invasions, facilitating or limiting colonization and spread of their avian hosts in new continents and islands. Several hypotheses could be applied to explain the role of parasites in avian invasions. The *Novel Weapon Hypotheses* (*NWH*) builds on the idea that invasive bird species gain advantages over native birds by bringing their own parasites to the new environments against which the introduced species but not the natives have evolved defenses. These co-introduced parasites may switch to native hosts and spread in the new communities, hence becoming themselves invasive parasites provoking serious damages to indigenous bird species. But the role of parasites in invasions may extend well beyond such direct effects, and hence some indirect species may also be expected. In this sense, the *Enemy Release Hypothesis (ERH)* poses that non-native bird species could become invaders because they have lost their co-evolved malaria parasites during the process of colonization, and thus they may increase their competitive ability and displace native species in the new areas. Conversely, the *Biotic Resistance Hypothesis* states that native parasites in the indigenous species may reduce the fitness of the potential bird invader and prevent its spread and establishment.

Malaria species are among the most pathogenic species of poultry and wildlife birds, being responsible of economic losses, mass mortality, population declines and even extinctions of many bird species worldwide after its introduction outside its native range. For all these reasons, the International Union for Conservation of Nature classifies avian *Plasmodium* to be among the 100 of the world's worst invasive alien species. In this review the authors will recapitulate the main scientific contributions showing the role of avian malaria parasites in the global spread of their bird hosts, evaluating the characteristics of both bird host and malaria parasite to become a fruitful invader. They will also highlight the extinctions and endangerment of numerous native species provoked by avian malaria invasions through the world (Hawaii, New Zealand, Galápagos Islands), and new menaces of avian malaria invasions in America.

Chapter 3 – Globally, biological invasion is a threat to economy, biodiversity, and human health. Most of the alien species were mediated by human activity, which is true for aquatic invasion species as well. China has become the world's second largest economic entity and the world's first aquaculture and fisheries trade powers. A total of 544 alien invasive species, including 51 of 100 of the world's worst invasive alien species, have successfully invaded China and have caused great damage to ecological environment, economic development, and public health; ultimately caused by the increase of trade, tourism and logistics. To date, however, few systematic reviews existed to summarize the process, impact and management measures concerning invasive species, especially for aquatic non-native invasive species in China. Here in this chapter, the authors first reviewed the advance of aquatic animal invasion, including the definition of aquatic animal invasion and their unique presence in the world biological invasion. Secondly, they listed 31 main invasive and potential nuisance aquatic alien animal species, 221introduced species including 100 aquaculture animals and 121 ornamental fish, summarized the ecological impact, economic loss and invasion progress, and gave some examples with respect to ecologic destruction. Thirdly, taking apple snail (*Pomacea canaliculata*) and fish as examples, the authors comprehensively analyzed the management strategy, and the current prevention and control measures for invasive alien aquatic animals. Finally, they gave some suggestions about the future management and research topics of aquatic animal invasion.

Chapter 4 – Natural grasslands are one of the ecosystems that have suffered the most intense series of human changes, even though they have not received enough conservation efforts. The proliferation of invasive shrubs and trees over grasslands and savannas around the world has been one of their severest modifications during the last century. The last remnants of South

American Pampas are not the exception, with a set of invasive plants of diverse origins thriving over native biodiversity. Among them, those producing animal-dispersed fruits represent a case of particular interest. St. Lucie's cherry (*Prunus mahaleb* L.) is a small fleshy-fruited tree native to Eurasia that invades natural grasslands in Southern Pampas, where the authors registered twenty animal species contributing to its dispersal.

In this chapter, the authors analyze temporal variations in flowering and availability of ripe fruits at the individual and population level. The species phenology in the area matches with reports from its native range, showing short flowering (19-22 days) and fruiting periods (27-34 days). Synchronicity in flowering and ripening of fruits are consistent with generalist strategies in plant-animal interactions, maximizing cross-pollination probabilities and the attraction of several generalist dispersal agents. The authors postulate that the lack of reproductive phenological overlapping with similar plants may play a key role in its invasion success. Their results suggest that the protection and shading of other trees might permit the fruits to remain in good condition in the canopies for the frugivores during longer periods. In early January, a massive fruit fall occurs and this might change the main dispersal role from birds consuming fruits in the canopies to mammals eating them directly from the floor, being the formers more related to short-distance dispersal and the latter possibly associated to long distance spread. Big trees with high density of fruits beneath the canopy might act as attraction foci to mammals and might have a key role in promoting long-distance dispersal. This suggests that part of the control efforts should be devoted to manage fruit availability in those attraction foci. *P. mahaleb* reproductive biology and its interactions with dispersal agents can explain to a considerable extent its invasive capacity in the study site and offer valuable information for its management. Incorporating these perspectives on mutualisms and invader-ecosystem interactions might improve the ability to predict invasion processes and contribute to better management decisions in other ecosystems invaded by similar fleshy-fruited species.

In: Biological Invasions
Editor: Rebecca Waterman

ISBN: 978-1-63482-019-6

Chapter 1

INVASIVE AQUATIC MICROORGANISMS: PATTERNS OF INTRODUCTION AND IMPACTS

Alessandro Saccà
Department of Biological and Environmental Sciences,
University of Messina, Messina, Italy

ABSTRACT

Alien species, defined as those species that are introduced to an area outside their natural range, are recognized as a major threat to the conservation of biodiversity and functioning of marine ecosystems. Coastal marine habitats are among the most heavily invaded systems on earth, which is derived, in part, by human-mediated transport of potentially invasive species in ships' ballast as well as intentional introductions for aquaculture and fisheries enhancement. Transfers of cultured fish and shellfish are also responsible for the accidental introduction of 'hitch-hiking' species, which could become invasive and severely affect the recipient ecosystem. In addition, man-induced climate change and eutrophication are expected to increase the susceptibility of ecosystems to invasion by alien species.

Although microorganisms are by far the most abundant and diverse of aquatic life forms, research on microbial alien species introductions has lagged behind that on metazoans and macroalgae. It is now evident, however, that due to human activities, thousands of species of marine viruses, prokaryotes and protists are being transported within and between oceans, irrespective of any natural barriers. Because of their metabolic plasticity, short generation times, wide tolerance to environmental conditions, and ability to form resting stages, microorganisms are likely the most successful invaders of coastal ecosystems.

It is well known that invasive microalgae capable of forming massive blooms can cause economic losses to aquaculture, fisheries and tourism activities and have important environmental and human health impacts. There is also growing evidence that, besides free-living species, pathogens and parasites are being frequently introduced into coastal systems, with potentially enormous consequences for economic activities and human health. Regulations at national and above-national levels are thus being implemented to face these emergent issues.

INTRODUCTION

Alien—also called non-indigenous or non-native—species are defined as those species that colonize an area beyond their natural range, where they reproduce and establish a population. It is known that plants, animals and microorganisms have been intentionally transferred with human discovery voyages for centuries and, possibly, since the shift from foraging to agricultural and pastoralist societies. As a result, many 'exotic' species are now among our preferred foods, dearest pets, good-looking houseplants and decorative aquarium weeds, but most of them cannot be considered alien species, as they grow well only in artificially-controlled conditions or in protected environments. Conversely, alien species are capable of ready acclimation to novel habitats, where they may find themselves unconstrained by the limiting factors—both abiotic and biotic—typical of their original habitat. In these 'favorable' conditions, they might outgrow, and ultimately overthrow, resident organisms with which they happen to compete: in this case, alien species are often dubbed 'invasive'. In fact, not all alien species that do become established turn out to be invasive—although some may become invasive after having been unremarkable for long time spans (Crooks and Soulè 1999).

Bioinvasions (invasions by alien species) occur through two processes: range expansions and introductions. Range expansions derive from dispersal by natural mechanisms, whereas introductions consist of transportation across natural barriers caused by human activity either directly or indirectly (Nehring 1998).

Many biological invasions are not a consequence of human interference: in fact, other animals accomplish active migrations and inadvertently move smaller organisms from their starting place all along the way to their destination. Moreover, herds of microbes, plants, and invertebrates have evolved adaptations to spread across distances and barriers even without the help of motile organisms, such as resting stages, seeds and other propagules, and whose comparative success has depended on these adaptations. Changes in ecological, environmental, or climatic conditions caused by human activity also may promote range expansion as well as facilitate introductions caused by human intervention.

Particularly interesting for their significant impact at the global scale are species introductions, but although estuarine and marine environments are known to be severely affected by this phenomenon, the study of non-indigenous species in these systems has lagged behind that of terrestrial and freshwater habitats, and can be dated back to about three decades ago (Carlton 1985; 1987). Since then, there has been increased interest in the invasion biology of coastal systems by macroorganisms, while microbial invasions have started to gain wider attention even more recently due to sometimes unapparent, yet important, consequences for human society and the environment.

In this chapter, I want to summarize the present knowledge on the invasions of aquatic habitats—particularly marine and brackish water—by microorganisms, with emphasis not only on their vectors and patterns of dissemination, but also on the ecological and evolutionary determinants behind their success. Through a few examples, I also wish to stress the environmental, social and economic impacts of microbial bioinvasions and, finally, briefly review the instruments and actions that have been advised to limit the phenomenon.

PATTERNS OF INVASION

Over the last centuries, with the development of increasingly efficient transport means and with the dramatic progress of human globalization, seaborne transfer rates of organisms have considerably intensified. Coastal zones are the start and end points of ships' movements and are also among the habitats most severely affected by anthropic impacts such as pollution, eutrophication, resource exploitation, and changes in sedimentation and in water circulation. As a consequence, coastal waters are among the most heavily invaded systems on earth.

Since humans started travelling by sea, wooden sailing ships were colonized by marine organisms fouling, boring into or nestling on the hull. It should not be surprising, then, if many of them—such as the marine mollusk *Teredo navalis,* which is specialized to bore through wood—now have cosmopolitan distributions. This transport means is typical for species capable of burrowing into or sticking firmly onto hard surfaces and withstanding high hydrodynamism—as well as for any species strictly associated with them, such as their parasites, pathogens and symbionts. Macroalgae and sessile macroinvertebrates are the most obvious organisms to take advantage of this means of dispersion.

Many sailing vessels, when voyaging, need to be ballasted to keep the craft upright. Large tankers and bulk cargo carriers use huge amounts of ballast water, which is often loaded in one region's coastal waters, for example in the port where cargo is unloaded, and discharged in a different region, where new cargo is stocked, often thousands of miles away. Discharged ballast water typically contains a variety of organisms, including species that can become invasive and cause extensive ecological, economic, and health consequences.

Indeed, ballast water discharges are alleged to be the major source of invasive macroorganisms in coastal waters of many countries and represent a suitable transport means also for free-living unicellular organisms. Thousands of different species of microbes are being transferred over long distances in ballast tanks at any one time (Carlton 1999). Most die out during travel due to the harsh conditions in ballast tanks; for others the port of release is an inhospitable habitat. Even when the environmental conditions are apparently suitable, many organisms fail to establish in the new environment due to negative ecological interactions with resident species, such as predation or competition. Nonetheless, cleaner ballast water, higher ships' speeds, and improved environmental management of ports, all favor worldwide dissemination of organisms by commercial and non-commercial shipping.

The majority of studies on organisms transported in ballast water have focused on metazoans (chiefly zooplankton) or on microalgae (mainly diatoms and dinoflagellates), although a variety of other microbial organisms are well documented (see e.g., Drake et al. 2001; and Drake et al. 2007). For example, a number of living protozoa were identified from water and sediments in the ballast tanks of cargo vessels arriving at Israeli ports in 1996, enabling the reconstruction of a multiplicity of interactions, including trophic relationships and host/parasite associations (Galil and Hulsmann 1997; Hulsmann and Galil 2001).

Bacteria and viruses can be transported in ballast tanks in many ways, for example in the ballast water itself, in sediments at the bottom of the tank or in biofilms covering the tank's walls. In addition, some bacterial and viral pathogens may occur—regularly or fortuitously—on the outer surfaces of invertebrates. It is worth mentioning that the bacterium *Vibrio cholerae* has been observed attached to the chitinous surfaces of copepods and other crustaceans (Huq et al. 2001; Lipp et al. 2002), to larval stages of planktonic invertebrates (Martinelli Filho et al. 2010), and to the mucilaginous envelopes of some phytoplankton (Huq et al. 2001; Lipp et al. 2002). Copepods and phytoplankton are common and often abundant in ballast water (*e.g.* Carlton and Geller 1993; Cohen 1998; Ruiz and Hines 1997; Subba Rao et al. 1994; MEPC 2003), and the association with these organisms might facilitate the transport of *V. cholerae* in ballast water and possibly its ability to successively infect a host (Colwell 1996; Lipp et al. 2002).

The fraction of microbes surviving the voyage in ballast waters depends on many factors, among which are the particular species or strain, the length of the journey, the water temperature, and so on (Alam et al. 2007; Gerba 2007; Gerba and Schaiberger 1975; Gerba and McLeod 1976; Lipson and Stotzky 1984; Sinclair et al. 2008). Theoretically, a ballast tank can act as an incubator with a consequent population growth of certain microbial species during travel: in the absence of light, photosynthesis will cease, determining the death of phytoplankton and causing, in turn, the starvation and death of zooplankton (e.g. Gollasch et al. 2000); the decomposition of these organisms may promote bacterial growth, resulting in increased viral replication (Drake et al. 2001), unless relatively high abundances of bacteria with antiviral mechanisms are present (Girones et al. 1989). Due to the complex relationships involved, however, it is possible that certain strains of bacteria or viruses will proliferate during travel in ballast tanks, even while the total numbers decline. Atypical microbial communities—and thus unusual interactions—may also occur in ballast tanks, given that supposedly empty tanks normally contain variable amounts of water and organisms and some mixing always occurs, bringing together species and strains from different habitats and regions. This can also provide the chance for horizontal gene transfer, with consequences at both ecological and evolutionary levels (see below).

Ballast water discharge is not the only cause of microbial marine species introductions. Numerous alien species of macroorganisms are deliberately transferred in previously uncolonized environments for fisheries and mariculture purposes. Associated with these, many inadvertently transported metazoans, macroalgae, and unicellular organisms can be found, among which parasites and pathogens of the intentionally imported species (Bax et al. 2003; Green et al. 2011; Oidtmann et al. 2011) as well as 'hitch-hiking' free-living organisms, which could become invasive and severely affect the recipient ecosystem (Guillemaud et al. 2011; Hulme 2005). Particularly, aquaculture practices involving the transport of live shellfish, especially oysters, have been identified as major vectors for alien species introductions in marine and estuarine systems (Elton 1958; Molnar et al. 2008). Indeed, fast airborne transport of shellfish in optimal and stable conditions is much more favorable for survival of 'hitch-hiking' species than longer maritime voyages in more variable environments such as ships' hulls or ballast tanks.

According to Carlton (2001), numerous other introduction means exist, namely those associated with drilling platforms, dry docks, navigation buoys and marina floats, amphibious planes and seaplanes, canals, public and private aquaria, movement and release for research and education purposes, floating marine debris, recreational equipment, and restoration initiatives. The impacts of these dispersion means taken separately is considered to be lower, with consequent scarcer interest by the scientific community, than the ones discussed above. This does not mean, however, that their effect is trivial, as a combination of vectors may actually be involved in many invasion events.

Invasions by Free-Living Microorganisms

Despite the fact that non-pathogenic microorganisms are by far the most abundant and diverse of aquatic life forms, contributing the majority of primary production and playing key roles in biogeochemical cycles of elements, their invasions are only recently receiving adequate attention. This is due, in part, to the obscure origin and difficult detection of such invasions (Desprez-Loustau et al. 2007; Wyatt and Carlton 2002), as opposed to spreads of pathogenic microbes, which often have obvious causes and dramatic impacts on conspicuous macroorganisms. Nevertheless, invasions of non-pathogenic microbes can have substantial effects on community composition and ecosystem functioning, including competitive exclusion of indigenous species, bottom-up effects on higher levels of the food web, depletion of nutrients, and so on.

Unfortunately, recognizing free-living invasive microbes is complicated by inherent difficulties. As shown by Gómez (2008), ascertaining phytoplankton invasions is challenging, due to problems in identification, taxonomic uncertainties and synonymies. Cryptic species, species complexes and polymorphic species are common among unicellular eukaryotes, while in prokaryotes, even applying the very concept of 'species' is often tricky (e.g. Staley 2006). Finally, establishing if a species has invaded an environment requires information on both previous and current community composition, but even so, until invaders attain relatively high abundances, they can easily go unnoticed (Bent and Forney 2008). In fact, at an early invasion stage potentially invasive species can be present at such low densities (Allee effect) that they are hardly detected, with consequent underestimation of invasion risk (Taylor and Hastings 2005).

Despite this, recorded invasions by non-pathogenic microorganisms are on the rise. In part, this is due to increased awareness of the phenomenon and to the subsequent proliferation of studies on the subject, but it is also favored by the recent introduction of high throughput molecular techniques targeting increasingly reliable DNA sequences, which have helped overcome many of the problems associated with the classical morphologic and molecular methods employed to describe microbial community composition (e.g. Mahé et al. 2014; Stoeck et al. 2010), although they are not exempt of flaws (see e.g., Lemos et al. 2011; Xiao et al. 2014).

But which microbes are the most likely to become invasive? Several studies have tried to outline an identikit of invasive species: being a generalist, having an r-strategy, a high phenotypic plasticity, and good dispersal capabilities are indicated as the most common traits among invasive macroorganisms (Sakai et al. 2001), and it appears reasonable to assume an important role for microorganisms too.

Invasive species may also display better performances than non-invaders. In fact, there is evidence suggesting that invasive free-living protists and cyanobacteria have relatively higher growth rates than native microorganisms of the same functional group (Istvanovics et al. 2000; Spaulding and Elwell 2007). Furthermore, although invasive microorganisms, might occupy largely similar ecological niches to those of resident species, they are often more efficient at a given function. In many environments there are resources that are not fully utilized or that could be exploited more efficiently if only alternative metabolic pathways were represented in the resident community. A microbial invader making use of locally non-represented, yet advantageous metabolic pathways is more likely to have success in these environments. Also mixotrophic organisms, which are able to exploit organic compounds in addition to performing photo- or chemosynthesis, are often favored over strictly autotrophs when inorganic nutrients or light are in short supply.

An invasive diatom *Didymosphenia geminata* (Bacillariophyceae), broadly diffused in low-nutrient habitats, is capable of utilizing organic phosphorus quite efficiently, resulting in a competitive advantage with respect to other phytoplankton (Whitton et al. 2009). Another invasive freshwater protist with worldwide distribution, *Gonyostomum semen* (Raphidiophyceae), causing nuisance algal blooms and known to produce allergic reactions in people swimming in lakes, is able to assimilate dissolved organic carbon and to migrate vertically, thus accessing resources in various zones of the water column (Rengefors et al. 2008; Salonen and Rosenberg 2000). *Cylindrospermopsis raciborskii*, an invasive cyanophycean that can be found in temperate lakes, avoids Nitrogen limitation by fixing atmospheric nitrogen (Briand et al. 2004) and can also assimilate phosphorus more efficiently than similar resident species (Istvanovics et al. 2000), resulting in a competitive advantage for nutrients. Finally, the ability of many microorganisms to withstand harsh environmental conditions (including those in ballast tanks) through resting stages (Ramette and Tiedje 2007) and to resume growth once they reach proper habitats or when the environmental conditions become suitable, could increase their invasive potential.

Invasive species, however, are not static entities: hybridization of alien species with native relatives has been shown to enhance their invasiveness (Lambrinos 2004), but also horizontal gene transfer, which is common in prokaryotes and also occurs in eukaryotic microbes, may give rise to evolutionary innovations that help microorganisms adapting to newly colonized environments (Breitbart and Rohwer 2005; Keeling and Palmer 2008; Ochman et al. 2000). Frequent gene transfer and high genetic diversity might indeed enable some microorganisms to become successful invaders (Hunt et al. 2008), but even invaded communities may benefit from evolutionary interactions with non-native species. In fact, invasive microbes may provide new genes for resident microbial communities through horizontal gene transfer, thus potentially increasing the resident community's genetic diversity (Mooney and Cleland 2001).

The recent, hot debate between supporters of the ‘cosmopolitanism’ and the ‘partial endemism’ paradigms in microbial biogeography (see e.g., Dolan 2005; Fenchel and Finlay 2004; Foissner 2006; Foissner et al. 2008) evidenced that, similar to macroorganisms, many microbial species do exhibit geographic patterns in their distributions at different scales (Martiny et al. 2006; Taylor et al. 2006; Telford et al. 2006; Vanormelingen et al. 2008; Winsborough et al.

2009; Whitaker et al. 2003). The greatly enhanced opportunities of dispersion provided by human activities, however, clearly give microorganisms which have not yet fully exploited their potential range the chance of becoming invaders. In this respect, species with poor dispersal abilities but with other characteristics increasing the probability of invasiveness can have unprecedented opportunities to reach and colonize novel habitats. This may be true not only for free-living microbes, but also for obligate symbionts, parasites and pathogens, whose spread opportunities are limited by the availability of hosts.

Considering its remarkably disjunct distribution (Saccà and Giuffrè 2014), the tintinnid ciliate *Rhizodomus tagatzi* appears as a perfect example of a species with poor dispersal abilities whose invasiveness has been greatly enhanced thanks to human activities. Probably native to the Gulf of Mexico, *R. tagatzi* has spread in several places around the northern hemisphere, most probably due to ballast water discharges as well as to bivalve transplants for aquaculture purposes (Saccà and Giuffrè 2014). Indeed, transfers of viable tintinnids via ballast waters have already been proposed (Chandrasekera and Fernando 2009; David et al., 2007; Pierce et al., 1997), but this mechanism alone does not completely explain the dispersion routes of this species. Interestingly, outside the Gulf of Mexico and the Caribbean Sea *R. tagatzi* has been found only in highly anthropized environments, and particularly in confined coastal features such as bays and lagoons converted into ports or exploited for aquaculture, where this ciliate can attain significantly high abundances in suitable conditions—including relatively elevated temperatures and high densities of nanophytoplankton—and notwithstanding severe hypoxia (Saccà et al. 2008).

Microbial Invasion Ecology

Community ecology theory has been used to understand biological invasions by introducing the concept of 'niche opportunity'. The latter describes the circumstances that promote invasions in terms of resources, natural enemies, and the abiotic environment, and can be considered as the potential provided by a given community for alien organisms to establish and develop large population sizes from initial low abundances. Successful invasions might thus depend on some favorable combination of resources availability, escape chances, and environmental conditions, including their fluctuations in time and space. Conversely, difficulty in introduced species being able to successfully invade an environment (invasion resistance) may be due to low levels of niche opportunities (Shea and Chesson 2002).

Elton had already assumed half a century ago (Elton 1958) that communities with high biodiversity should show relatively elevated invasion resistance. Recent models and some experimental studies indeed suggest that there is a negative relationship between invasion success and species diversity (Knops et al. 1999; Levine and D'Antonio 1999; Naeem et al. 2000; Stachowicz et al. 1999). Nevertheless, there are circumstances in which the traits of resident and invading species are the major factors influencing vulnerability to invasion (Lavorel et al. 1999; McGrady-Steed et al. 1997).

Indeed, the structure and dynamics of biotic interactions in an ecosystem clearly affect the success of potentially invasive species. In fact, a recent invader might not be threatened by resident highly specialized pathogens, parasites or predators and then gain a considerable advantage by escaping from its own specialist natural enemies or losing them early in the invasion process when its density is too low to maintain their populations (Settle and Wilson, 1990; Torchin et al. 1996). It is known, for example, that pathogens often contribute to the control of a species in its native range, while their absence in a newly invaded environment can facilitate the initial success of an invasion (Mitchell et al. 2006).

Pathogens, parasites, specialized grazers or predators not present at an early stage of the invasion, however, may arrive later (acting in turn as invaders) reestablishing the control of the invasive species. This happened to *Alexandrium minutum*, which invaded the Atlantic coastal waters off France, producing uncontrolled blooms. Two decades later, this toxic, marine dinoflagellate was effectively restrained by a parasite and the blooms never reappeared (Chambouvet et al. 2008; 2011). This is in accordance with the proposed enemy release hypothesis (ERH), also referred to as the herbivore escape, predator escape or ecological release hypothesis (Crawley 2009; Keane and Crawley 2002; Williamson 1996). The ERH asserts that non-indigenous species, on introduction to a new environment, should experience a drop in regulation by natural enemies, resulting in an improved population dynamics. This may play a significant role in the spread and subsequent control of invasive microorganisms.

Similarly, pathogens and parasites usually have a greater impact on populations that have never been exposed to them (naive host populations), even if they have little or no pathogenic effect on related host organisms in their native regions. This typically occurs because the naive host lacks defenses, such as an effective immunologic response, or other adaptations (e.g.

behavioral) which can reduce the incidence or severity of the pathogenic impact. A greater impact on naive hosts can also occur because pathogens may have evolved to be less harmful to host populations with which they have been in contact for a long time, thus enhancing their own long-term persistence (Anderson and May 1982; Fenner and Myers 1977).

Global climate change could act synergistically with other anthropogenic stressors, such as eutrophication or overfishing, to promote microbial invasions (Walther et al. 2009): global warming could favor invasions by tropical and subtropical microbes into temperate latitudes (Occhipinti and Ambrogi 2007), while anthropogenic alterations in resident community structure, may enhance niche opportunity. For example, free-living tropical toxic cyanobacteria are currently spreading into temperate latitudes, particularly where ecosystem integrity is endangered by human activities. In truth, since heat waves appear to stimulate emergence of many marine microbes from dormant cysts (e.g. Wiedner et al. 2007), predicted increases in frequency of such extreme events (Walther et al. 2009) are likely to promote microbial invasions where conditions are favorable.

Impacts of Invasions

Biological invasions by macroorganisms are considered the second major cause of biodiversity loss, after habitat destruction (Vitousek et al. 1997). The negative impact of microbial invasions on biodiversity is presumably equally important, although assessing it is much less straightforward. Invasions by alien microbes can also have serious effects on ecosystem functioning and services. For example, phytoplankton, unicellular organisms performing photosynthesis in the water column, are the main primary producers in aquatic ecosystems and thus have the potential to exert bottom-up control over the entire community as well as to directly influence the biogeochemical cycles of elements. For this reason, when invasive microalgae form blooms and become highly dominant over the usual resident species they may affect the entire microbial food web, and hence the whole ecosystem. Furthermore, among phytoplankton species are those which can produce and release secondary metabolites that negatively affect the growth of other organisms, a phenomenon known as allelopathy (Fistarol et al. 2003; 2004; Graneli and Hansen 2006). These can be true toxins that may exert their effects on various species, including humans.

The transport of toxic dinoflagellates in ballast tanks has received particular attention. Approximately 60 marine dinoflagellates are known to produce substances that are toxic and sometimes fatal to humans or animals (Doblin and Dobbs 2006). Some of these form cysts that can remain dormant and viable for months or even years, and such cysts have been found in the sediments in ballast tanks at up to 22,500 per gram of wet sediment (Hallegraeff and Bolch 1992; Hamer et al. 2000; 2001). There is good evidence that some toxic dinoflagellates have been introduced into new regions of the world in ballast discharges (Dobbs and Rogerson 2005; Drake et al. 2007; Hallegraeff 1998; Hallegraeff et al. 1995; Lilly et al. 2001; McMinn et al. 1997). *Pfiesteria piscicida* and *P. shumwayae*, dinoflagellates that have been implicated in numerous fish kills, were found in one of four vessels sampled on arrival in Chesapeake Bay or the Great Lakes (Doblin et al. 2004; Drake et al. 2005) and in about 10% of the tanks sampled for residual ballast water on NOBOB vessels (vessels that declare 'no ballast on board' but that typically have relatively small volumes of ballast water remaining in their tanks) arriving at ports in the Great Lakes in 2001-2002 (Johengen et al. 2005). Remarkable cases of algal blooms are the so-called 'red tides', spectacular increases of the biomass of a single (or few) autotrophic or mixotrophic microbial species (either prokaryote or eukaryote), to such extent that the water assumes the color of the main algal photosynthetic pigments (from pale pink to brownish). This phenomenon typically occurs in summer, prompted by higher temperature and light availability and is often coupled to eutrophication or other significant changes in the nutrient balance. As mentioned above, introduced alien species, not controlled by their natural enemies and possibly more efficient in utilizing limiting resources than residents, can take advantage of such conditions and become invasive, often aided by favorable hydrologic features (Tester and Steidinger 1997). Certain red tides are associated with production of natural toxins, depletion of dissolved oxygen or other detrimental effects, and rank among the so-called harmful algal blooms (HABs). These kinds of red tides are often correlated with mortalities of fish, birds, marine mammals, and other organisms, and an increase in their frequency and magnitude at worldwide scale has been observed (Anderson 1989; Anderson et al. 2002; Hallegraeff 1993; Smayda 1989; 1990; GEOHAB 2001).

Karenia brevis (also known as *Gymnodinium Breve*), formerly endemic in the Gulf of Mexico and subsequently introduced in Japanese waters and in the Mediterranean Sea, is known for causing 'red tides' which have significant consequences for human health and ecosystem functioning, as well as serious economic costs to fishery, aquaculture and tourism. This mixotrophic dinoflagellate produces a family of neurotoxins called brevetoxins, long known for being responsible for massive fish kills and a human nonlethal form of food intoxication, neurotoxic shellfish poisoning (NSP), due to its accumulation in the tissues of wild as well as of cultured bivalves (McFarren et al. 1965; Shimizu et al. 1986). Adverse effects on human health of the inhalation of air-borne toxins have also been known for about a century (Pierce and Henry 2008; Taylor 1917), while a direct relationship between *K. brevis* blooms and mortality of marine mammals, sea turtles and sea birds has been demonstrated more recently (Brand et al. 2012; Landsberg 2002; Walsh et al. 2010; Watkins et al. 2008). Further, as the conditions sustaining the bloom of *K. brevis* deteriorate, mass mortality and subsequent organic matter decomposition cause severe oxygen depletion, with possible long-term effects on benthic communities (Pierce and Henry 2008).

Another dinoflagellate, *Prorocentrum minimum*, is widely distributed in temperate and subtropical waters where it causes HABs in many coastal and estuarine environments. This protist has experienced rapid geographical expansion in the last half century, and huge blooms have been reported from many coastal sites around the world (see Heil et al. 2005 for a review). A relationship appears to exist between the increased frequency of *P. minimum* blooms and the escalation of coastal eutrophication phenomena (e.g. Glibert et al. 2001; Marasović 1986; Marasović et al. 1990; Kondo et al. 1990b). In fact, this species shows a series of physiologic adaptations that may confer a competitive advantage over local phytoplankton populations upon its introduction into nutrient-enriched environments: it is capable of rapid growth, it can use a variety of inorganic and organic nutrient substrates, it is adapted to a wide range of irradiance conditions, salinities, and temperatures, and also shows potential allelopathic properties.

The spread of *P. minimum* is particularly well documented in the Black Sea, the Baltic Sea, and the saline Lake Nakanoumi in Japan (Bjergskov et al. 1990; Hajdu et al. 2000; Hansen 2002; Kondo et al. 1990a,b,c; Marasović et al. 1990; Moncheva et al. 1995; Olesen 2001; Petrova-Karadjova 1984), but there is also evidence of its expansion elsewhere in the world (including the Adriatic Sea and the coasts of the Philippines, Mexico, and China). Possible factors responsible for expansion and increase in bloom frequency include transport in ballast water, aquaculture activities, and eutrophication (GEOHAB 2001). Indeed, vegetative cells of *P. minimum* have been retrieved from cargo ships ballast tanks in the Pacific Ocean (Yoshida et al. 1996) and the northern Baltic Sea (Olenin et al. 2000).

Although *P. minimum* is generally considered not to be toxic to marine invertebrates, larger blooms have been reported to cause adverse impacts due to high algal biomass and related effects, among which localized oxygen depletion, pH change, and significant light attenuation. Harmful consequences associated with *P. minimum* blooms range from fish and zoobenthic mortalities to shellfish aquaculture die-offs, attributable to both indirect (e.g. oxygen depletion following biomass decomposition) and direct toxic effects. *P. minimum* blooms also appear to be potentially harmful to humans via shellfish poisoning (Tangen 1983; Taylor et al. 1995).

Among potentially invasive microalgae, it is not only dinoflagellates that can have negative impacts. The diatom *Coscinodiscus wailesii*, supposedly native to the Indian and Pacific Oceans, has been introduced to European waters and was first detected in the English Channel in 1977. Climate-driven changes in ocean circulation might have facilitated its invasion of many coastal shelf zones in northern Europe (Edwards et al. 2001), while its tolerance to a wide range of environmental conditions have probably favored its establishment and rapid spread. At present, it can be found along the Atlantic coast off France, around the British Isles, and in the North Sea (Laing and Gollasch 2002), but also along the cold temperate coasts of the western side of the Atlantic Ocean, on the Peruvian shelf, and scattered along Korean and Japanese coasts (OBIS 2014). *C. wailesii* is considered a potentially harmful species due to its production of insoluble mucilage which has adverse effects on fishing activities (Edwards et al. 2001). It may also develop high biomasses leading to oxygen and nutrient depletion, with severe impacts on mussel and seaweed cultivation (Fernandes et al. 2001; Proença and Fernandes 2004). For example, during bloom events, the depletion of inorganic nitrogen in areas cultivated with the seaweed *Porphyra* (Nori) causes thallus whitening and consequent economic loss. (Nishikawa and Hori 2004; Nishikawa and Yamaguchi 2008).

Besides free-living microorganisms, there is also growing evidence that pathogens and parasites are being frequently introduced into coastal systems, with potentially enormous consequences for economic activities and human health. In general, our knowledge of the diseases of marine and estuarine organisms is quite limited. Much of what we do know is derived from disease outbreaks in animals kept in captivity or used in aquaculture, and much less is known about the etiology of diseases occurring in wild populations.

Disease outbreaks among cultivated mollusks species are particularly well documented for oysters, and some of them have probably involved introductions of pathogens into new regions as a result of the transfer of these mollusks for aquaculture enhancement purposes. For example, disease epidemics affecting cultivated populations of the 'Japanese oyster' *Crassostrea gigas* in European countries since 2008 are most probably due to oyster imports from the northwestern Pacific (Mineur et al. 2014a). In France and in other European countries this species has been occasionally affected by a strain of ostreid herpesviruses (*Malacoherpesviridae*) since the early 1990s (OsHV-1 'reference'), but starting in 2008 a new, more virulent variant has emerged (OsHV-1 μVar), which is presently affecting the European oyster industry. A study by Mineur et al. (2014b) suggests that these viruses most probably have an Asian origin and that their introduction in European farming sites has occurred thanks to imports of *C. gigas* from the native area in East Asia intended to maintain or improve stock diversity.

Among oyster pathogens likely spread through mollusks transplants, two haplosporidian protozoa, *Bonamia ostreae* and *Haplosporidium nelsoni*, have been introduced respectively in Europe (Cigarria and Elston 1997; Goulletquer et al. 2002; Van Banning 1987), and in California, eastern North America and Europe (Andrews 1984; Burreson and Ford 2004; Burreson et al. 2000; Ford and Haskin 1982; Goulletquer et al. 2002; Renault et al. 2000), even though they might have been transferred in ballast waters as well (e.g. Burreson and Ford 2004). A third pathogenic protozoan, *Perkinsus marinus,* (Perkinsozoa) has been likely introduced in the northeastern US via oyster movements (Ford 1996).

One of the most studied fish pathogens is the viral hemorrhagic septicemia virus (VHSV), a rhabdovirus responsible for extensive losses of cultured and wild fish, which has expanded its geographic range in the last two decades. Initially, it was considered to be limited to freshwater fish in Western Europe (Wolf, 1988), but it was later identified in Pacific salmon (Batts et al. 1993; Winton et al. 1999), as well as in seawater-farmed species, such as rainbow trout (Dale et al. 2009). Then it was found in over 50 freshwater and marine fish species in the northern hemisphere with variable virulence (Gagné et al. 2007; Kent et al. 1998; Meyers and Winton 1995). The occurrence of massive fish mortality events starting in 2005 in the Great Lakes led to the hypothesis that VHSV is a recent invader, possibly introduced via ship ballast water (Carlton and Geller 1993; Lumsden 2005; Sieracky et al. 2014). Although ballast water has been implicated in most alien species introductions in the Great Lakes (Grigorovich et al. 2003; Holeck et al. 2004; Mills et al. 1993), however, there are currently no data to support a key role of ships in the introduction or dispersal of VHSV (Bain et al. 2010). Nevertheless, ship decontamination and port monitoring are being implemented (e.g. Cangelosi and Mays 2006) to prevent invasions by non-native species, including microorganisms.

As of human pathogens, the already mentioned bacterium, *Vibrio cholerae*, is the etiologic agent of human cholera. The world has experienced seven pandemics of cholera over the last two centuries, the seventh of which began in Indonesia in 1961. From that date and until the 1980s, it spread in chronological order through Southeast Asia, India, Africa, and the Middle East (Colwell 1996; Den Enden Erwin 2004). In the 1990s it appeared in coastal areas of Peru and then spread throughout Southern and Central America. The same strain that caused the latter outbreak was detected by the US Food and Drug Administration (USFDA) in fish and oysters in Mobile Bay, Alabama (DePaola et al. 1992; CDC 1993) and, concurrently, in the ballast water of about one-third of the examined ships arriving in the US from ports in Latin America (McCarthy and Khambaty 1994; McCarthy et al. 1992). It was concluded that the South American pathogenic strain of *V. cholerae* was introduced to US Gulf Coast waters by cargo vessels, probably in ballast water, and that cargo vessels may also have initially introduced the strain in South America from Asia. While a wide consensus exists regarding the introduction of this strain to US coasts by ballast waters (Dobbs and Rogerson 2005; Drake et al. 2001; 2007; Tibbetts 1997), its first arrival in South America appears less clear, with both endorsements of the ballast water inception (Aguirre-Macedo et al. 2008; Anderson 1991; Ditchfield 1993; Epstein 1995; Epstein et al. 1993; Tibbetts 1997) and opposite opinions (Colwell 1996; Lipp et al. 2002).

Control Measures

As discussed above, invasive marine species can pose serious threats to ecosystem services, economic activities, and, not least, human health. Scientific and technologic advances can be of help in mitigating the issue, but are not decisive. For example, the improvement in anti-fouling paint technology has generally decreased the amount of fouling on the hulls of cruise ships and merchant vessels, but sensitive spots such as rudders, gratings, holes, and sea chests remain difficult to paint and thus subject to high levels of fouling, preventing ships from being entirely free from fouling organisms.

Specific regulatory programs to prevent adverse impacts of bioinvasions have been developed, but they have mainly focused on international shipping and ballast water (see e.g., Cohen et al. 2010). Even so, the slow progress of international initiatives and the lack of any global monitoring system have led to the independent development of national regulations in many countries. Two examples are the Australian Ballast Water Management Requirements and the Brazilian Resolution RDC 217, both introduced in 2001.

As for the USA, since 1997 the National Invasive Species Act has required vessels arriving in U.S. ports to submit specific Ballast Water Reporting Forms. The US Coast Guard and the National Ballast Information Clearinghouse (NBIC), established in 1999, gather information on the transport and discharge of ballast water from such forms on a national scale, and the NBIC uses these data to estimate the rate of ballast water exchange and the release of foreign ballast water, and reports periodically to the US Congress. The Environmental Protection Agency and the US Coast Guard are also developing initiatives to regulate the concentration of living organisms discharged in the ballast water of ships in order to minimize the spread of non-indigenous species in US waterways.

Successful prevention and management of the threats due to marine invasive species is only possible, however, if handled at a global and regional level. Biological invasions have recently become a key priority for European authorities, which are currently involved in developing new regulations (EC 2013*a*), after recognizing the failure of the present 'fragmented and incoherent policy set up at EU and national levels' (EC 2013*b*). The International Maritime Organization (IMO) introduced voluntary ballast water guidelines in 1997 and three years later it joined forces with the Global Environment Facility (GEF), the United Nations Development Programme (UNDP), member governments and the shipping industry, resulting in the GloBallast project, a program aimed at assisting developing countries to implement effective measures to reduce the risk of transfer of alien marine species in ballast water. The International Convention for the Control and Management of Ships' Ballast Water and Sediments (BWM) adopted in 2004 aims to prevent the spread of harmful aquatic organisms between different regions, through the introduction of standards and procedures according to a ship-specific ballast water management plan. The convention establishes that all ships have to carry a ballast water record book and an international ballast water management certificate, and that most ships will be required to install an on-board ballast water treatment system.

Several techniques have been advised for the treatment of ballast water, including heat treatment, chemical treatment, ultrafiltration, ultraviolet light, and so on. None of these methods, however, is exempt from drawbacks, or is appropriate for treating thousands of tons of ballast water in a short time.

While waiting for a suitable solution, ballast water exchange at sea is recommended, consisting of pumping through the ballast tank an amount of oceanic seawater three times its volume. This is expected to lead to approximately 95% exchange of the original ballast water, replacing it with oceanic water, which poses little risk to coastal ecosystems. The efficacy of this procedure, however, is unknown, since a number of microbes will remain in the tank, both in the residual water and in the sediments.

Another problem is that ballast water treatment systems (BWTS) are advised only for aquatic organisms in the size classes of 10–50 μm and > 50 μm. Organisms smaller than 10 μm are not considered when checking BWTS according to IMO standards. This size class includes viruses, bacteria, and many microalgae and protozoa. In fact, it is known that the great majority of all phytoplankton organisms are smaller than 10 μm, and that many harmful phytoplankton species belong to this size class.

New standards and techniques of ballast water treatment are therefore needed, bearing in mind the necessity for trade-offs between both the environmental and the economic costs and benefits. Thus an overall risk management framework is necessary to discriminate between vessels posing the highest threat, and hence requiring treatment and deserving a high monitoring level, and comparatively low-risk vessels that can be allowed to discharge their untreated ballast water and be subject to a lower level of monitoring. Similarly, the benefits and costs of prevention, control, and management of alien marine species invasions should be considered in view of the process of social and economic development of a region, that is to say, its sustainable development.

While it is widely accepted that prevention is more effective than control, it is also clear that no single vector accounts for all invasive species introductions. Consequently, prevention actions must be focused on multiple vectors, in the framework of a comprehensive management system (GESAMP 2007). Many microbes, especially free-living forms, have a multitude of dispersal pathways that are poorly known and, therefore, virtually impossible to control at present (Vellinga et al. 2009). Initiatives such as New Zealand's Biodiversity Strategy and Australia's proposed National System for the Prevention and Management of Introduced Marine Pests take a more inclusive approach to target marine-introduced species and represent a first step towards multiple issue focus of bioinvasions control. These initiatives include a range of prevention measures covering a variety of vectors and sources, monitoring activities to detect new introductions or the spread of existing invaders, coordinated emergency procedures in response to new invasions, and plans for long-term control of established non-native species. Even so, very little is being done at the international level to reduce the threat from the many other vectors than ballast waters. This is a major weakness in the current global response to marine invasive species. Finally, not only should unintentional introductions be addressed, but also international instruments that regulate purposeful introductions and translocations of non-native species are needed (Naylor et al. 2001; Shine et al. 2000). In this respect, preventing the movement of plants or animals across regions and applying quarantine measures might be effective in reducing the spread of some microbial taxa, such as 'hitch-hiking' free-living species associated with aquaculture organisms.

For example, following a first European Directive issued in 1991, transfers of shellfish have been regulated by European legislation, which consent all movements inside Europe, except for possible restrictions during disease outbreaks (EC 1991; 1995; 2003; 2006). Although imports from outside Europe are subject to severe restrictions, the current legislation still allows the importation of live mollusks from some Pacific regions (e.g., from the west coast of North America) for aquaculture purposes (EC 2004*a*; 2008; 2009), and from others (e.g., from Japan and South Korea) for food marketing (EC 2004*b*; SANCO 2013).

In conclusion, as suggested by Bax et al. (2003), management of marine invasive species is best treated as an epidemiologic problem. This implies that policy design should account for three main points: first, prevention is more effective (and cheaper) than control; second, all of the important vectors should be identified in order to efficiently prevent alien species from spreading into new areas; third, reducing the risk of future invasions and managing extant invasions is a more reasonable aim than absolute avoidance of the invasion of a species. Anyway, as we learn more about the ecological traits and patterns of spread of invasive microbes, new approaches for controlling this global-scale issue will likely be developed.

REFERENCES

Aguirre-Macedo, M. L., Vidal-Martinez, V. M., Herrera-Silveira, J. A., Valdés-Lozano, D. S., Herrera-Rodríguez, M. and Olvera-Novoa, M. A. (2008). Ballast water as a vector of coral pathogens in the Gulf of Mexico: The case of the Cayo Arcas coral reef. *Mar. Poll. Bull., 56*, 1570-1577.

Alam, A., Sultana, M., Nair, G. B., Siddique, A. K., Hasan, N. A., Sack, R. B., Sack, D. A., Ahmed, K., Sadique, U. A., Watanabe, H., Grim, C. J., Huq, A. and Colwell, R. R. (2007). Viable but nonculturable Vibrio cholerae O1 in biofilms in the aquatic environment and their role in cholera transmission. *PNAS, 104*, 17801-17806.

Anderson, C. (1991). Cholera epidemic traced to risk miscalculation. *Nature, 354*, 255.

Anderson, D. M. (1989). Toxic algal bloom and red tides: a global perspective. In T. Okaichi, D. M. Anderson and T. Nemoto (Eds.), *Red Tides: Biology, Environmental Science and Technology* (pp. 11-16). New York, NY, USA: Elsevier.

Anderson, D. M., Glibert, P. M. and Burkholder, J. M. (2002). Harmful algal blooms and eutrophication: nutrient sources, composition and consequences. *Estuaries, 25*, 562–584.

Anderson, R. M. and May, R. M. (1982). Coevolution of hosts and parasites. *Parasitology, 85*, 411-426.

Andrews, J. D. (1984). Epizootiology of diseases of oysters (*Crassostrea virginica*), and parasites of associated organisms in eastern North America. *Helgoland Marine Research, 37,* 149-166.

Bain, M. B., Cornwell, E. R., Hope, K. M., Eckerlin, G. E., Casey, R. N., Groocock, G. H., Getchell, R. G., Bowser, P. R., Winton, J. R., Batts, W. N., Cangelosi, A. and Casey, J. W. (2010). Distribution of an invasive aquatic pathogen (viral hemorrhagic septicemia virus) in the Great Lakes and its relationship to shipping. *PloS one*, *5*, e10156.

Batts, W. N., Arakawa, C. K., Bernard, J. and Winton, J. R. (1993). Isolates of viral hemorrhagic septicemia virus from North America and Europe can be detected and distinguished by DNA probes. *Dis. Aquat. Org., 17*, 61-71.

Bax, N., Williamson, A., Aguero, M., Gonzalez, E., and Geeves, W. (2003). Marine invasive alien species: a threat to global biodiversity. *Marine policy*, *27*, 313-323.

Bent, S. J. and Forney, L. J. (2008). The tragedy of the uncommon: understanding limitations in the analysis of microbial diversity. *ISME J., 2,* 689-695.

Bjergskov, T., Larsen, J., Moestrup, Ø., Sørensen, H. M. and Krogh, P. (1990) Toksiske of Potentielt Toksiske Alger i Danske Farvande. The Fish Inspection Service, Danish Ministry of Fisheries, Copenhagen (in Danish), 200 pp.

Boudouresque, C. F., and Verlaque, M. (2002). Biological pollution in the Mediterranean Sea: invasive versus introduced macrophytes. *Mar. Poll. Bull.*, *44*, 32-38.

Brand, L. E., Campbell, L. and Bresnan, E. (2012). *Karenia*: the biology and ecology of a toxic genus. *Harmful Algae 14,* 156-178.

Breitbart, M. and Rohwer, F. (2005). Here a virus, there a virus, everywhere the same virus? *Trends Microbiol., 13*, 278-284.

Burreson, E. M. and Ford, S. E. (2004). A review of recent information on the Haplosporidia, with special reference to *Haplosporidium nelsoni* (MSX disease). *Aquatic Living Resources, 17*, 499-517.

Burreson, E. M., Stokes, N. A. and Friedman, C. S. (2000). Increased virulence in an introduced pathogen: *Haplosporidium nelsoni* (MSX) in the eastern oyster *Crassostrea virginica*. *J. Aquat. Anim. Health, 12*, 1-8.

Cangelosi, A., and Mays, N. (2006). Great ships for the Great Lakes? Commercial vessels free of invasive species in the Great Lakes–St. Lawrence Seaway System. A scoping report for the Great Ships Initiative. Northeast-Midwest Institute, Washington DC 143 pp.

Carlton, J. T. (1987). Patterns of transoceanic marine biological invasions in the Pacific Ocean. *Bull. Mar. Sci., 41*, 452-465.

Carlton, J. T. (1999). The scale and ecological consequences of biological invasions in the world's oceans. In: O. T. Sandlund, P. J. Schei and A. Viken (Eds.), *Invasive species and biodiversity management.* (pp. 195-212). Dordrecht, The Nethrlands: Kluwer Academic Publishers.

Carlton, J. T. (2001). Introduced species in US coastal waters: environmental impacts and management priorities. Arlington, Virginia, United States: Pew Oceans Commission Report. 28 pp.

Carlton, J. T. and Geller, J. B. (1993). Ecological roulette: The global transport of nonindigenous marine organisms. *Science, 261,* 78-82.

Carlton, J.T. (1985). Transoceanic and interoceanic dispersal of coastal marine organisms: the biology of ballast water. *Oceanogr. Mar. Biol. Annu. Rev., 23*, 313-374.

CDC (1993). Isolation of *Vibrio cholerae* O1 from Oysters - Mobile Bay, 1991-1992. *Centers for Disease Control, Morbidity and Mortality Weekly Report, 42*, 91-93.

Chambouvet, A., Laabir, M., Sengco, M., Vaquer, A., and Guillou, L. (2011). Genetic diversity of Amoebophryidae (Syndiniales) during *Alexandrium catenella/tamarense* (Dinophyceae) blooms in the Thau lagoon (Mediterranean Sea, France). *Research in microbiology*, *162*, 959-968.

Chambouvet, A., Morin, P., Marie, D. and Guillou, L. (2008). Control of toxic marine dinoflagellate blooms by serial parasitic killers. *Science, 322*, 1254-1257.

Chandrasekera, W. U. and Fernando, M. A. S. T. (2009) Accidental introduction of alien plankton into the Sri Lankan coastal zone through ballast water of cargo ships. Sri Lanka. *J. Aquat. Sci., 14*, 87-103.

Cigarria, J. and Elston, R. (1997). Independent introduction of *Bonamia ostreae,* a parasite of *Ostrea edulis,* to Spain. *Dis. Aquat. Org., 29*, 157-158.

Cohen, A. N. (1998). Ships' Ballast Water and the Introduction of Exotic Organisms into the San Francisco Estuary: Current Status of the Problem and Options for Management. A report for CALFED and the California Urban Water Agencies. San Francisco Estuary Institute, Richmond CA.

Cohen, A.N. 2010. Non-native Bacterial and Viral Pathogens in Ballast Water: Potential for Impacts to ESA-listed Species under NOAA's Jurisdiction. A report prepared for the National Oceanic and Atmospheric Administration, National Marine Fisheries Service, Endangered Species Division, Silver Spring, MD. Center for Research on Aquatic Bioinvasions (CRAB), Richmond, CA. 85 pp.

Colwell, R. R. (1996). Global climate and infectious disease: the cholera paradigm. *Science 274,* 2025-2031.

Crawley, M. J. (2009). Plant-Herbivore Dynamics, in M. J. Crawley (Ed.), *Plant Ecology*, Second Edition, (pp. 401-474). Oxford, UK: Blackwell Publishing Ltd.

Crooks, J. A. and Soulè, M. E. (1999). Lag times in population explosions of invasive species: causes and implications. In O. T. Sandlund, P. J. Schei and A. Viken (Eds.), *Invasive species and biodiversity management.* (p.p. 103–125). Dordrecht, The Netherlands: Kluwer Academic Publishers.

Dale, O. B., Orpetveit, I., Lyngstad, T. M., Kahns, S., Skall, H. F., Olesen, N. J., and Dannevig, B. H. (2009). Outbreak of viral haemorrhagic septicaemia (VHS) in seawater-farmed rainbow trout in Norway caused by VHS virus Genotype III. *Dis. Aquat. Org., 85*, 93-103.

David, M., Gollasch, S., Cabrini, M., Perkovič, M., Bošnjak, D., and Virgilio, D. (2007). Results from the first ballast water sampling study in the Mediterranean Sea–the Port of Koper study. *Mar. Poll. Bull., 54*, 53-65.

Den Enden Erwin, V. (2004). Illustrated Lecture Notes on Tropical Medicine, Edition 2004. Institute of Tropical Medicine, Antwerp. [Online] http://itg.content-e.eu/Generated/pubx/173/cholera/epidemiology.htm.

DePaola, A., Capers, G. M., Motes, M. L., Olsvik, O., Fields, P. I., Wells, J., Wachsmuth, I. K., Cebula, T. A., Koch, W. H., Khambaty, F., Payne, W. L. and Wentz, B. A. (1992). Isolation of Latin American epidemic strain of *Vibrio cholerae* O1 from US Gulf Coast. *Lancet, 339*, 624-625.

Desprez-Loustau, M. L., Robin, C., Buee, M., Courtecuisse, R., Garbaye, J., Suffert, F., Sache, I. and Rizzo, D. M. (2007). The fungal dimension of biological invasions. *Trends Ecol. Evol., 22,* 472-480.

Dobbs, F. C. and Rogerson, A. (2005). Ridding ships' ballast water of microorganisms. *Env. Sci. Technol. 39*, 259A-264A.

Doblin, M. A. and Dobbs, F. C. (2006). Setting a size-exclusion limit to remove or inactivate toxic dinoflagellate cysts in ships' ballast water. *Mar. Poll. Bull., 52,* 259-263.

Doblin, M. A., Drake, L. A., Coyne, K. J., Rublee, P. A. and Dobbs, F. C. (2004). *Pfiesteria* species identified in ships' ballast water and residuals: a possible vector for introductions to coastal areas. in K. A. Steidinger, J. H. Landsberg, C. R. Tomas and G. A. Vargo (Eds.), *Harmful Algae* (pp. 317-319), Florida Fish and Wildlife Conservation Commission, Florida Institute of Oceanography, and Intergovernmental Oceanographic Commission of UNESCO.

Dolan, J. R. (2005). An introduction to the biogeography of aquatic microbes. *Aquat. Microb. Ecol., 41,* 39-48.

Drake, L. A., Choi, K. H., Ruiz, G. M. and Dobbs, F. C. (2001). Global redistribution of bacterioplankton and virioplankton communities. *Biological Invasions, 3*, 193-199.

Drake, L. A., Choi, K.-H., Ruiz, G. M. and Dobbs, F. C. (2001). Global redistribution of bacterioplankton and virioplankton communities. *Biological Invasions, 3*, 193-199.

Drake, L. A., Doblin, M. A. and Dobbs, F. C. (2007). Potential microbial bioinvasions via ships' ballast water, sediment, and biofilm. *Mar. Pollut. Bull., 55*, 333-341.

Drake, L. A., Meyer, A. E., Forsberg, R. L., Baier, R. E., Doblin, M. A., Heinemann, S., Johnson, W. P., Koch, M., Rublee, P. A. and Dobbs, F. C (2005). Potential invasion of microorganisms and pathogens via 'interior hull fouling': biofilms inside ballast-water tanks. *Biological Invasions, 7*, 969-982.

EC (European Community). (1991). Council directive of 28 January 1991 concerning the animal health conditions governing the placing on the market of aquaculture animals and products (91/67/EEC). *Official Journal of the European Union, L46,* 1-18.

EC (European Community). (1995). Council directive of 22 December 1995 introducing minimum Community measures for the control of certain diseases affecting bivalve mollusks (95/70/EC). *Official Journal of the European Union, L332,* 33-44.

EC (European Community). (2003). Commission decision of 14 November 2003 laying down the animal health conditions and certification requirements for imports of mollusks, their eggs and gametes for further growth, fattening, relaying of human consumption (2003/804/EC). *Official Journal of the European Union, L302,* 22-33.

EC (European Community). (2004*a*). Commission decision (2004/319/EC) of 30 March 2004 amending Annex I to Decision 2003/804/EC laying down the animal health conditions and certification requirements for imports of mollusks, their eggs and gametes for further growth, fattening, relaying or human consumption. *Official Journal of the European Union, L102,* 73-74.

EC (European Community). (2004*b*). Regulation (EC) No 854/2004 of the European Parliament and of the council of 29 April 2004 laying down specific rules for the organization of official controls on products of animal origin intended for human consumption. *Official Journal of the European Union, L155,* 206-320.

EC (European Community). (2006). Council directive (2006/88/EC) of 24 october 2006 on animal health requirements for aquaculture animals and products thereof, and on the prevention and control of certain diseases in aquatic animals. *Official Journal of the European Union, L328,* 14-56.

EC (European Community). (2008). Commission regulation (EC) No 1251/2008 of 12 December 2008 implementing Council Directive 2006/88/EC as regards conditions and certification requirements for the placing on the market and the import into the Community of aquaculture animals and products thereof and laying down a list of vector species. *Official Journal of the European Union, L337,* 41-75.

EC (European Community). (2009). Commission regulation (EC) No 719/2009 of 6 August 2009 amending Regulation No 1251/2008 as regards the list of third countries and territories from which certain crustaceans and ornamental aquatic animals may be imported into the Community. *Official Journal of the European Union, L205,* 10-14.

EC (European Community). (2013*a*). Proposal for a regulation of the European Parliament and of the Council on the prevention and management of the introduction and spread of invasive alien species (* COM/2013/0620 final–2013/0307 (COD)). Available from http://eur-lex.europa.eu/legal-content/EN/TXT/PDF/?uri=CELEX:52013PC0620&from=EN (accessed 22 November 2014).

EC (European Community). (2013*b*). Commission staff working document, executive summary of the impact assessment accompanying the document "Proposal for a regulation of the European Parliament and of the Council on the prevention and management of the introduction and spread of invasive alien species". Available from http://register.consilium.europa.eu/doc/srv?l=EN&t=PDF&f=ST+13457+2013+ADD+1 (accessed 22 November 2014).

Edwards, M., John, A. W. G., Johns, D. G. and Reid, P. C. (2001). Case history and persistence of the non-indigenous diatom *Coscinodiscus wailesii* in the north-east Atlantic. *J. Mar. Biol. Ass. U. K., 81*, 207-211.

Elton, C. S. (1958). *The Ecology of Invasions by Animals and Plants*. London, UK: Methuen Publishing Ltd. 181 pp.

Fenchel, T. and Finlay, B. J. (2004). The ubiquity of small species: patterns of local and global diversity. *BioScience, 54,* 777-784.

Fenner, F. and Myers, K. (1977). Myxoma virus and myxomatosis in retrospect: the first quarter century of a new disease. In: E. Kurstak and K. Maramorosch (Eds.), *Viruses and Environment* (pp. 539-570). New York, USA: Academic Press.

Fernandes, L., Zehnder-Alves, L. and Bassfeld, J. (2001). The recently established diatom *Coscinodiscus wailesii* (Coscinodiscales, Bacillariophyta) in Brazilian waters. I: Remarks on morphology and distribution. *Phycol. Res., 49,* 89-96.

Fistarol, G. O., Legrand, C., and Granéli, E. (2003). Allelopathic effect of Prymnesium parvum on a natural plankton community. *Mar. ecol. Prog. Ser., 255*, 115-125.

Foissner, W. (2006). Biogeography and dispersal of micro-organisms: a review emphasizing protists. *Acta Protozool., 45,* 111-136.

Foissner, W., Chao, A. and Katz, L. A. (2008). Diversity and geographic distribution of ciliates (Protista: Ciliophora). *Biodivers. Conserv., 17,* 345-363.

Ford, S. E. (1996). Range extension by the oyster parasite *Perkinsus marinus* into the northeastern United States: response to climate change? *J. Shellfish Res., 15*, 45-56.

Ford, S. E. and Haskin H. H. (1982). History and epizootiology of *Haplosporidium nelsoni* (MSX), an oyster pathogen in Delaware Bay, 1957-1980. *J. Invertebr. Pathol., 40,* 118-141.

Gagné, N., MacKinnon, A. M., Boston, L., Souter, B., Cook-Versloot, M., Griffiths, S. and Olivier, G. (2007). Isolation of viral haemorrhagic septicaemia virus from mummichog, stickleback, striped bass and brown trout in eastern Canada. *J. Fish Dis., 30,* 213-223.

Galil, B. S. and Hülsmann, N. (1997). Protist transport via ballast water—biological classification of ballast tanks by food web interactions. *Eur. J. Protistol., 33*, 244-253.

GEOHAB - Global Ecology and Oceanography of Harmful Algal Blooms Programme. (2001). In P. Glibert and G. Pitcher (Eds.), *Science Plan* (pp. 86). Baltimore, MD, and Paris, France: SCOR and IOC.

Gerba, C. P. (2007). Virus occurrence and survival in the environmental waters. In W. A. Bosch (Ed.), *Human Viruses in Water* (pp. 91-108). Amsterdam, The Netherlands: Elsevier.

Gerba, C. P. and McLeod, J. S. (1976). Effect of sediments on the survival of Escherichia coli in marine waters. *Appl. Environ. Microbiol., 32,* 114-120.

Gerba, C. P. and Schaiberger, G. E. (1975). Effect of particulates on virus survival in seawater. *J. Water Poll. Control Fed., 47,* 93-103.

GESAMP (2002). 'Treatment and management of ships' ballast water to control introductions of non-indigenous species', Report of the GESAMP Ballast Water Correspondence Group to Marine Environment and Protection Committee of the International Maritime Organization. *MEPC 48/2, 2002.*

Girones, R., Jofre, J. T. and Bosch, A. (1989). Isolation of marine bacteria with antiviral properties. *Can. J. Microbiol., 35,* 1015-1021.

Glibert, P.M., Magnien, R., Lomas, M.W., Alexander, J., Fan, C., Haramoto. E., Trice, M. and Kana, T.M. (2001). Harmful algal blooms in the Chesapeake and Coastal Bays of Maryland, USA: comparison of 1997, 1998, and 1999 events. *Estuaries, 24,* 875-883.

Gollasch, S., Lenz, J., Dammer, M. and Andres, H.G. (2000). Survival of tropical ballast water organisms during a cruise from the Indian Ocean to the North Sea. J. Plankton Res., 22, 923-937.

Gómez, F. (2008). Phytoplankton invasions: Comments on the validity of categorizing the non-indigenous dinoflagellates and diatoms in European Seas. *Mar. Pollut. Bull., 56*, 620-628.

Goulletquer, P., Bachelet, G., Sauriau, P. G. and Noel, P. (2002). Open Atlantic coast of Europe - a century of introduced species into French waters. In: E. Leppakoski, S. Gollasch and S. Olenin (Eds.), *Invasive Aquatic Species of Europe Distribution Impacts and Management* (pp. 276-290). Dordrecht, The Netherlands: Springer.

Granéli, E. and Hansen, P. J. (2006). Allelopathy in Harmful Algae: A Mechanism to Compete for Resources? In E. Granéli, and J.T. Turner (Eds.), *Ecology of Harmful Algae. Ecological Studies, Vol. 189* (pp. 189-201). Berlin, Heidelberg, Germany: Springer Verlag.

Green, D. M., M. Werkman, L. A. Munro, R. R. Kao, I. Z. Kiss, and L. Danon. 2011. Tools to study trends in community structure: application to fish and livestock trading networks. *Preventive Veterinary Medecine 99,* 225–228.

Grigorovich, I. A., Colautti, R. I., Mills, E. L., Holeck, K., Ballert, A. G., and MacIsaac, H. J. (2003). Ballast-mediated animal introductions in the Laurentian Great Lakes: retrospective and prospective analyses. *Can. J. Fish Aquat. Sci., 60*, 740-756.

Guillemaud, T., Ciosi, M., Lombaert, E. and Estoup. A. (2011). Biological invasions in agricultural settings; insights from evolutionary biology and population genetics. *Compte Rendus Biologies, 334,* 237-246.

Hajdu, S., Edler, L., Olenina, I. and Witek, B. (2000). Spreading and establishment of the potentially toxic dinoflagellate *Prorocentrum minimum* in the Baltic Sea. *Intern. Rev. Hydrobiol., 85,* 561-575.

Hallegraeff, G. M. (1993). A review of harmful algal blooms and their apparent global increase. *Phycologia, 32,* 79-99.

Hallegraeff, G. M. (1998). Transport of toxic dinoflagellates via ships' ballast water: bioeconomic risk assessment and efficacy of possible ballast water management strategies. *Mar. Ecol. Prog. Ser., 168,* 297-309.

Hallegraeff, G. M. and C. J. Bolch. (1992). Transport of diatom and dinoflagellate resting spores in ships' ballast water: implications for plankton biogeography and aquaculture. *J. Plankton Res., 14,* 1067-1084.

Hallegraeff, G. M., McCausland, M. A. and Brown, R. K. (1995). Early warning of toxic dinoflagellate blooms of Gymnodinium catenatum in southern Tasmanian waters. *J. Plankton Res., 17,* 1163-1176.

Hamer, J. P., Lucas, I. A. N. and McCollin, T. A. (2001). Harmful dinoflagellate resting cysts in ships' ballast tank sediments: potential for introduction into English and Welsh waters. *Phycologia, 40,* 246-255.

Hamer, J. P., McCollin, T. A. and Lucas, I. A. N. (2000). Dinoflagellate cysts in ballast tank sediments: between tank variability. *Mar. Poll. Bull., 40,* 731-733.

Hansen, P. J. (2002). The effect of high pH on the growth and survival of marine phytoplankton: implications for species succession. *Aquat. Microb. Ecol., 28,* 279-288.

Heil, C. A., Glibert, P. M., and Fan, C. (2005). *Prorocentrum minimum* (Pavillard) Schiller: A review of a harmful algal bloom species of growing worldwide importance. *Harmful Algae*, *4*, 449-470.

Holeck, K. T., Mills, E. L., MacIsaac, H. J., Dochoda, M. R., Colautti, R. I. and Ricciardi, A. (2004) Bridging troubled waters: Biological invasions, transoceanic shipping, and the Laurentian Great Lakes. *BioScience, 54,* 919-929.

Hulme, P. E. (2005). Nursery crimes: agriculture as victim and perpetrator in the spread of invasive species. In *Crop Science and Technology* (pp. 733-740). Alton, United Kingdom: British Crop Protection Council.

Hulsmann, N. and Galil, B. S. (2001). The effects of freshwater flushing on marine heterotrophic protists - implications for ballast water management. *Mar. Poll. Bull., 42,* 1082-1086.

Hunt, D. E., David, L. A., Gevers, D., Preheim, S. P., Alm, E. J. and Polz, M. F. (2008). Resource partitioning and sympatric differentiation among closely related bacterioplankton. *Science, 320,* 1081-1085.

Huq, A., Sack, R. B. and Colwell, R. R. (2001). Cholera and global ecosystems. In J. L. Aron and J. Patz (Eds.), *Ecosystem Change and Public Health: A Global Perspective* (pp. 327-347). Baltimore, MD: Johns Hopkins University Press.

Istvanovics, V., Shafik, H. M., Presing, M. and Juhos, S. (2000). Growth and phosphate uptake kinetics of the cyanobacterium, *Cylindrospermopsis raciborskii* (Cyanophyceae) in throughflow cultures. *Freshw. Biol., 43,* 257-275.

Johengen, T. J., Reid, D., Fahnenstiel, G., MacIsaac, H., Dobbs, F. C., Doblin, M. A. and Jenkins, P. T. (2005). Assessment of transoceanic NOBOB vessels and low-salinity ballast water as vectors for nonindigenous species introductions to the Great Lakes. Final report for the project to the Great Lakes Protection Fund, the National Oceanic and Atmospheric Administration, the U.S. Environmental Protection, and the U.S. Coast Guard.

Keane, R. M., and Crawley, M. J. (2002). Exotic plant invasions and the enemy release hypothesis. *Trends Ecol. Evol.*, *17*, 164-170.

Keeling, P. J. and Palmer, J. D. (2008). Horizontal gene transfer in eukaryotic evolution. *Nat. Rev. Gen., 9,* 605-618.

Kent, M. L., Traxler, G. S., Kieser, D., Richard, J., Dawe, S. C., Shaw, R. W., Prosperi-Porta, G., Ketcheson, J. and Evelyn, T. P. T. (1998) Survey of Salmonid Pathogens in Ocean-Caught Fishes in British Columbia, Canada. *J. Aquat. Anim. Health, 10,* 211-219.

Knops, J. M. H., Tilman, D., Haddad, N. M., Naeem, S., Mitchell, C. E., Haarstad, J., Ritchie, M. E., Howe, K. M., Reich, P. B., Siemann, E. and Groth, J. (1999). Effects of plant species richness on invasions dynamics, disease outbreaks, insect abundances, and diversity. *Ecol. Lett., 2,* 286.293.

Kondo, K., Seike, Y. S., Date, Y. (1990a). Red tides in the brackish Lake Nakanoumi. 1: The frequency and causative species of red tides. *Bull. Plankton Soc. Japan, 36,* 103-110.

Kondo, K., Seike, Y. S., Date, Y. (1990b). Red tides in the brackish Lake Nakanoumi. 2: Relationships between the occurrence of *Prorocentrum minimum* red tide and environmental conditions. *Bull. Plankt. Soc. Japan, 37,* 19-34.

Kondo, K., Seike, Y. S., Date, Y. (1990c). Red tides in the brackish Lake Nakanoumi. 3: The stimulative effects of organic substances in the interstitial water of bottom sediments and in the excreta from *Skeletonema costatum* on the growth of *Prorocentrum minimum. Bull. Plankt. Soc. Japan, 37,* 35-47.

Laing, I., and Gollasch, S. (2002). *Coscinodiscus wailesii*—a nuisance diatom in European waters. In E. Leppäkoski, S. Gollasch and S. Olenin (Eds.), *Invasive Aquatic Species of Europe. Distribution, Impacts and Management* (pp. 53-55). Dordrecht, The Netherlands: Springer.

Lambrinos, J. G. (2004). How interactions between ecology and evolution influence contemporary invasion dynamics. *Ecology, 85*, 2061-2070.

Landsberg, J. H. (2002). The effects of harmful algal blooms on aquatic organisms. *Rev. Fish. Sci., 10,* 113-390.

Lavorel, S., Prieur-Richard, A. H. and Grigulis, K. (1999). Invasibility and diversity of plant communities: from patterns to processes. *Diversity Distrib., 5,* 41-49.

Lemos, L. N., Fulthorpe, R. R., Triplett, E. W. and Roesch, L. F. (2011). Rethinking microbial diversity analysis in the high throughput sequencing era. *J. Microbiol. Methods*, *86*, 42-51.

Levine, J. M. and D'Antonio, C. M. (1999). Elton revisited: a review of evidence linking diversity and invasibility. *Oikos, 87,* 15-26.

Lilly, E. L., Kulis, D. M., Gentien, P. and Anderson, D. M. (2001). Paralytic shellfish poisoning toxins in France linked to a human-introduced strain of *Alexandrium catenella* from the western Pacific: evidence from DNA and toxin analysis. *J. Plankton Res., 24,* 443-452.

Lipp, E. K., Huq, A. and Colwell, R. R. (2002). Effects of global climate on infectious disease: the cholera model. *Clin. Microbiol. Rev. 15,* 757-770.

Lipson, S. M. and Stotzky, G. (1984). Effect of proteins on reovirus adsorption to clay minerals. *Appl. Environ. Microbiol. 48,* 525-530.

Lumsden, J. S. (2005). A mortality event in freshwater drum (*Aplodinotus grunniens*) from Lake Ontario, associated with viral hemorrhagic septicemia virus (VHSV), type IV. *Wildlife Health Centre Newsletter, 11,* 10.

Mahé, F., Mayor, J., Bunge, J., Chi, J., Siemensmeyer, T., Stoeck, T., Wahl, B., Paprotka, T., Filker, S. and Dunthorn, M. (2014). Comparing High-throughput Platforms for Sequencing the V4 Region of SSU-rDNA in Environmental Microbial Eukaryotic Diversity Surveys. *J. Euk. Microbiol.*, DOI: 10.1111/jeu.12187

Marasović, I., Pucher-Petkovic, T. and Petrova-Karadjova, V. (1990). *Prorocentrum minimum* (Dinophyceae) in the Adriatic and Black Sea. *J. Mar. Biol. Assoc. UK, 70,* 473-476.

Marasović, I. (1986). Occurrence of *Prorocentrum minimum* in the Adriatic Sea. *Rapp. Comm. Int. Mer Médit., 30*, 186.

Marasović, I., Pucher-Petković, T., Petrova-Karadjova, V. (1990). *Prorocentrum minimum* (Dinophyceae) in the Adriatic and Black Sea. *J. Mar. Biol. Assoc. UK, 70,* 473-476.

Martinelli Filho, J. E., Lopes, R. M., Rivera, I. N. C. and Colwell, R. R. (2010). *Vibrio cholerae* O1 detection in estuarine and coastal zooplankton. *J. Plankton Res.*, doi:10.1093/plankt/fbq093.

Martiny, J. B. H., Bohannan, B. J. M., Brown, J. H., Colwell, R. K., Fuhrman, J. A., Green, J. L., Horner-Devine, M. C., Kane, M., Adams Krumins, J., Kuske, C. R., Morin, P. J., Naeem, S., Øvreås, L., Reysenbach, A.-L., Smith, V. H. and Staley, J. T. (2006). Microbial biogeography: putting microorganisms on the map. *Nat. Rev. Microbiol., 4,* 102-112.

McCarthy, S. A. and Khambaty, F. M. (1994). International dissemination of epidemic *Vibrio cholerae* by cargo ship ballast and other nonpotable waters. *Appl. Environ. Microbiol., 60,* 2597-2601.

McCarthy, S. A., Gaines, J. L., McPhearson, R. M. and Guarino, A. M. (1992). Toxigenic *Vibrio cholerae* O1 and cargo ships entering Gulf of Mexico. *Lancet, 339,* 624-625.

McFarren, E. F., Tanabe, H., Silva, F. J., Wilson, W. B., Campbell, J. E., and Lewis, K. H. (1965). The occurrence of a ciguatera-like poison in oysters, clams, and *Gymnodinium breve* cultures. *Toxicon, 3*, 111-123.

McGrady-Steed, J., Harris, P. M. and Morin, P. J. (1997) Biodiversity regulates ecosystem predictability. *Nature, 390,* 162-165.

McMinn, A., Hallegraeff, G. M., Thomson, P., Jenkinson, A. V. and Heijnis, H. (1997). Cyst and radionucleotide evidence for the recent introduction of the toxic dinoflagellate *Gymnodinium catenatum* into Tasmanian waters. *Mar. Ecol. Prog. Ser., 161,* 165-172.

MEPC (2003). Harmful Aquatic Organisms in Ballast Water: Comments on Draft regulation E-2: Concentrations of Organisms Delivered in Ships' Ballast Water in the Absence of any Treatment: Establishing a Baseline for Consideration of Treatment Efficacy. MEPC 49/2/21, 49th Session, Agenda Item 2, 23 May 2003. Marine Environment Protection Committee, International Maritime Organization, London, UK.

Meyers, T. R. and Winton, J. R. (1995). Viral hemorrhagic septicemia virus in North America. *Ann. Rev. Fish. Dis., 5,* 3-24.

Mills, E. L., Leach, J. H., Carlton, J. T. and Secor, C. L. (1993). Exotic species in the Great Lakes: A history of biotic crises and anthropogenic introductions. *J. Great Lakes Res., 19,* 1-54.

Mineur, F., Le Roux, A., Maggs, C. A. and Verlaque, M. (2014a). Positive feedback loop between introductions of non-indigenous marine species and cultivation of oysters in Europe. *Conserv. Biol., 28,* 1667-1676

Mineur, F., Provan, J., and Arnott, G. (2014b). Phylogeographical analyses of shellfish viruses: inferring a geographical origin for ostreid herpesviruses OsHV-1 (Malacoherpesviridae). *Marine Biology*, 1-12.

Mitchell, C. E., Agrawal, A. A., Bever, J. D., Gilbert, G. S., Hufbauer, R. A., Klironomos, J. N. Maron, J. L., Morris, W. F., Parker, I. M., Power,A. G., Seabloom, E. W., Torchin, M. E. and Vázquez, D. P. (2006). Biotic interactions and plant invasions. *Ecol. Lett., 9,* 726-740.

Molnar, J. L., R. L. Gamboa, C. Revenga, and Spalding, M. D. (2008). Assessing the global threat of invasive species to marine biodiversity. *Front. Ecol. Environ., 6,* 485-492.

Moncheva, D., Petrova-Kardjova, V., Palasov, A. (1995). Harmful algal blooms along the Bulgarian sea coast and possible patterns of fish and zoobenthos mortalities. In P. Lassus, G. Arzul, E. Erard-Le Denn, P. Gentien, C. Marcaillou-Le Baut (Eds.), *Harmful Marine Algal Blooms* (pp. 193-198). Paris, France: Lavoisier.

Mooney, H. A. and Cleland, E. E. (2001). The evolutionary impact of invasive species. *Proc. Natl Acad. Sci. USA, 98,* 5446-5451.

Naeem, S., Knops, J. M., Tilman, D., Howe, K. M., Kennedy, T., and Gale, S. (2000). Plant diversity increases resistance to invasion in the absence of covarying extrinsic factors. *Oikos*, *91*, 97-108.

Naylor, R. L., Williams, S. L., and Strong, D. R. (2001). Aquaculture – a gateway for exotic species. *Science, **294***, 1655-56.

Nehring, S. (1998). Non-indigenous phytoplankton species in the North Sea: supposed region of origin and possible transport vector. *Arch. Fish. Mar. Res.*, *46*, 181-194.

Nishikawa, T. and Yamaguchi, M. (2008). Effect of temperature on light-limited growth of the harmful diatom Coscinodiscus wailesii, a causative organism in the bleaching of aquacultured *Porphyra* thalli. *Harmful Algae, 7,* 561-566.

Nishikawa, T., and Hori, Y. (2004). Effects of nitrogen, phosphorus and silicon on a growth of a diatom Coscinodiscus wailesii causing *Porphyra* bleaching isolated from Harima-Nada, Seto Inland Sea, Japan. *Nippon Suisan Gakkaishi, 70,* 872-878.

OBIS (2014). Data from the Ocean Biogeographic Information System. Intergovernmental Oceanographic Commission of UNESCO. Web. http://www.iobis.org (consulted 11 November 2014).

Occhipinti-Ambrogi, A. (2007). Global change and marine communities: alien species and climate change. *Mar. Poll. Bull.*, *55*, 342-352.

Ochman, H., Lawrence, J. G. and Groisman, E. A. (2000). Lateral gene transfer and the nature of bacterial innovation. *Nature, 405,* 299-304.

Oidtmann, B. C., Trush, M. A., Denham, K. L. and Peeler, E. J. (2011). International and national biosecurity strategies in aquatic animal health. *Aquaculture, 320,* 22-33.

Olenin, S., Gollasch, S., Jonušas, S., and Rimkutė, I. (2000). En-Route Investigations of Plankton in Ballast Water on a Ship's Voyage from the Baltic Sea to the Open Atlantic Coast of Europe. *Int. Rev. Hydrobiol.*, *85*, 577-596.

Olesen, M. (2001). Sedimentation in Mariager Fjord, Denmark: the impact of sinking velocity on system productivity. *Ophelia 55,* 11-26.

Petrova-Karadjova, V. J. (1984). Changes in plankton flora in Bulgarian Black Sea waters under the influence of eutrophication. *Proceedings of the Institute of Fisheries, Varna 21,* 105-112.

Pierce, R. H. and Henry, M. S. (2008). Harmful algal toxins of the Florida red tide (*Karenia brevis*): natural chemical stressors in South Florida coastal ecosystems. *Ecotoxicology, 17,* 623–631.

Pierce, R. W., Carlton, J. T., Carlton, D. A. and Geller, J. B. (1997). Ballast water as a vector for tintinnid transport. *Mar. Ecol. Prog. Ser., 149,* 295-297.

Proença, L. A. O. and Fernandes, L. F. (2004). Introdução de microalgas no ambiente marinho: impactos negativos e fatores controladores. In: J. S. V. Silva and, R. C. C. L. Souza (Eds.). *Água de lastro e bioinvasão* (pp. 77-97). Rio de Janeiro, Brazil: Interciência.

Ramette, A. and Tiedje, J. M. (2007). Biogeography: an emerging cornerstone for understanding prokaryotic diversity, ecology, and evolution. *Microb. Ecol., 53,* 197-207.

Renault, T., Stokes, N. A., Chollet, B., Cochennec, N., Berthe, F., Gérard, A. and Burreson, E. M. (2000). Haplosporidiosis in the Pacific oyster *Crassostrea gigas* from the French Atlantic coast. *Dis. Aquat. Org., 42,* 207-214.

Rengefors, K., Palsson, C., Hansson, L.A. and Heiberg, L. (2008). Cell lysis of competitors and osmotrophy enhance growth of the bloom-forming alga Gonyostomum semen. *Aquat. Microb. Ecol., 51,* 87-96.

Ruiz, G. M. and Hines, A. H. (1997). The Risk of Nonindigenous Species Invasion in Prince William Sound Associated with Oil Tanker Traffic and Ballast Water Management: Pilot Study. Report to Regional Citizen's Advisory Committee of Prince William Sound, Valdez AK.

Saccà, A., and Giuffrè, G. (2013). Biogeography and ecology of *Rhizodomus tagatzi*, a presumptive invasive tintinnid ciliate. *J. Plankton Res.*, *35*, 894-906.

Saccà, A., Guglielmo, L., and Bruni, V. (2008). Vertical and temporal microbial community patterns in a meromictic coastal lake influenced by the Straits of Messina upwelling system. *Hydrobiologia*, *600*, 89-104.

Sakai, A. K., Allendorf, F. W., Holt, J. S., Lodge, D. M., Molofsky, J., With, K. A. et al. (2001). The population biology of invasive species. *Annu. Rev. Ecol. Syst., 32,* 305-332.

Salonen, K. and Rosenberg, M. (2000). Advantages from diel vertical migration can explain the dominance of Gonyostomum semen (Raphidophyceae) in a small, steeply-stratified humic lake. *J. Plankton Res., 22,* 1841-1853.

SANCO (Directorate-General for Health and Consumer Protection). (2013). Listing of third country establishments after Regulation 854/2004. Available from http://ec.europa.eu/food/food/biosafety/establishments /third_country/new_estab_lists_en.htm (accessed 22 November 2014).

Settle, W. H. and Wilson, L. T. (1990). Invasion by the variegated leafhopper and biotic interactions: parasitism, competition and apparent competition. *Ecology, 71,* 1461-1470.

Shea, K., and Chesson, P. (2002). Community ecology theory as a framework for biological invasions. *Trends Ecol. Evol.*, *17*, 170-176.

Shimizu, Y., Chou, H. N., Bando, H., Van Duyne, G., and Clardy, J. (1986). Structure of brevetoxin A (GB-1 toxin), the most potent toxin in the Florida red tide organism *Gymnodinium breve* (*Ptychodiscus brevis*). *J. Am. Chem. Soc.*, *108*, 514-515.

Shine, C., Williams, N., and Gündling, L. (2000). Environmental Policy and Law Paper. Vol 40. Gland, Switzerland: IUCN.

Sieracki, J. L., Bossenbroek, J. M. and Faisal, M. (2014). Modeling the secondary spread of viral hemorrhagic septicemia virus (VHSV) by commercial shipping in the Laurentian Great Lakes. *Biological Invasions*, *16*, 1043-1053.

Sinclair, R., Boone, S. A., Greenberg, D., Keim, P. and Gerba, C. P. (2008). Persistence of Category A Select Agents in the Environment, *Appl. Environ. Microbiol., 74,* 555-563.

Smayda, T. J. (1989). Primary production and the global epidemic of phytoplankton blooms in the sea: a linkage? In E. M. Cosper, V. M. Bricelj and E. J. Carpenter (Eds.), *Novel Phytoplankton Blooms Coastal and Estuarine Studies* No. 35. (pp. 449-484). New York, USA: Springer-Verlag.

Smayda, T. J. (1990). Novel and nuisance phytoplankton blooms in the sea: evidence for a global epidemic. In E. Granéli, B. Sundstrom, L. Edler and D. M. Anderson (Eds.), *Toxic Marine Phytoplankton* (pp. 29-40). New York, USA: Elsevier.

Spaulding, S. and Elwell, L. (2007). Increase in Nuisance Blooms and Geographic Expansion of the Freshwater Diatom *Didymosphenia geminata*. U.S. Geological Survey Report, 2007-1425, 38 pp.

Stachowicz, J. J., Whitlatch, R. B., and Osman, R. W. (1999). Species diversity and invasion resistance in a marine ecosystem. *Science*, *286,* 1577-1579.

Staley, J. T. (2006). The bacterial species dilemma and the genomic–phylogenetic species concept. *Philos. Trans. R. Soc. Lond., B, Biol. Sci., 361*, 1899-1909.

Stoeck, T., Bass, D., Nebel, M., Christen, R., Jones, M. D., Breiner, H. W. and Richards, T. A. (2010). Multiple marker parallel tag environmental DNA sequencing reveals a highly complex eukaryotic community in marine anoxic water. *Molecular Ecology*, *19*(s1), 21-31.

Subba Rao, D. V., Sprules, W. G., Locke, A. and Carlton, J. T. (1994). Exotic phytoplankton from ships' ballast waters: Risk of potential spread to mariculture sites on Canada's East Coast. *Can. Data Rep. Fish. Aquatic Sci., 937,* 1-51.

Tangen, K. (1983). Shellfish poisoning and the occurrence of potentially toxic dinoflagellates in Norwegian waters. *Sarsia, 68*, 1-7.

Taylor, C. M. and Hastings, A. (2005). Allee effects in biological invasions. *Ecol. Lett., 8,* 895-908.

Taylor, F. J. R., Fukuyo, Y. and Larsen, J. (1995). Taxonomy of harmful dinoflagellates. In G. M. Hallegraeff, D. M. Anderson and A. D. Cembella (Eds.), *Manual on Harmful Marine Microalgae. IOC Manuals and Guides No. 33* (pp. 33). Intergovernmental Oceanographic Commission of UNESCO, Paris.

Taylor, H. (1917). Mortality of fishes on the west Coast of Florida. Report of the US Commissioner of Fisheries. Bureau of Fisheries Document 848. Government Printing Office, Washington, DC.

Taylor, J. W., Turner, E., Townsend, J. P., Dettman, J. R. and Jacobson, D. (2006). Eukaryotic microbes, species recognition and the geographic limits of species: examples from the kingdom Fungi. *Philos. Trans. R. Soc. Lond., B, Biol. Sci., 361,* 1947-1963.

Telford, R. J., Vandvik, V. and Birks, H. J. B. (2006). Dispersal limitations matter for microbial morphospecies. *Science, 312,* 1015.

Tester, P. A. and Steidinger, K. A. (1997). *Gymnodinium breve* red tide blooms: Initiation, transport, and consequences of surface circulation. *Limnol. Oceanogr., 42,* 1039-1051.

Tibbetts, J. (1997). Exotic invasion. *Environ. Health Perspect., 105,* 590-593.

Torchin, M. E., Lafferty, K. D. and Kuris, A. M. (1996). Infestation of an introduced host, the European green crab, *Carcinus maenas*, by a symbiotic nemertean egg predator, *Carcinonemertes epialti. J. Parasitol., 82,* 449-453.

Van Banning, P. (1987). Further results of the *Bonamia ostreae* challenge tests in Dutch oyster culture. *Aquaculture, 67,* 191-194.

Vanormelingen, P., Verleyen, E. and Vyverman, W. (2008). The diversity and distribution of diatoms: from cosmopolitanism to narrow endemism. *Biodivers. Conserv., 17,* 393-405.

Vellinga, E. C., Wolfe, B. E. and Pringle, A. (2009). Global patterns of ectomycorrhizal introductions. *New Phytol., 181,* 960-973.

Vitousek, P. M., D'Antonio, C. M., Loope, L. L., Rejmanek, M. and Westbrooks, R. (1997). Introduced species: a significant component of human-caused global change. *New Zealand J. Ecol., 21,* 1-16.

Walsh, C. J., Leggett, S. R., Carter, B. J. and Colle, C. (2010). Effects of brevetoxin exposure on the immune system of loggerhead sea turtles. *Aquat. Toxicol., 97,* 293-303.

Walther, G. R., Roques, A., Hulme, P. E., Sykes, M. T., Pysek, P., Kuhn, I. et al. (2009). Alien species in a warmer world: risks and opportunities. *Trends Ecol. Evol., 24,* 686-693.

Watkins, S. M., Reich, A., Fleming, L. E. and Hammond, R. (2008). Neurotoxic shellfish poisoning. *Marine Drugs 6,* 431-455.

Whitaker, R. J., Grogan, D. W. and Taylor, J. W. (2003). Geographic barriers isolate endemic populations of hyperthermophilic archaea. *Science, 301,* 976-978.

Whitton, B. A., Ellwood, N. T. W. and Kawecka, B. (2009). Biology of the freshwater diatom *Didymosphenia*: a review. *Hydrobiologia, 630,* 1-37.

Wiedner, C., Rücker, J., Brüggemann, R. and Nixdorf, B. (2007). Climate change affects timing and size of populations of an invasive cyanobacterium in temperate regions. *Oecologia*, *152*, 473-484.

Williamson, M. (1996) *Biological Invasions*, London, UK: Chapman & Hall. 256 pp.

Winsborough, B. M., Theriot, E. and Czarnecki, D. B. (2009). Diatoms on a continental ''island'': Lazarus species, marine disjuncts and other endemic diatoms of the Cuatro Cienegas basin, Coahuila, Mexico. *Nova Hedwigia, Suppl., 135,* 257-274.

Winton, J. R., Batts, W. N., Nishizawa, T. (1989). Characterization of the first North American isolates of viral hemorrhagic septicemia virus. *Am. Fish. Soc. Fish. Health News, 17.* 2-3.

Wolf, K. (1988). Viral hemorrhagic septicemia. In K. Wolf (Ed.), *Fish viruses and fish viral diseases* (pp 217-249). Ithaca, NY: Cornell University Press.

Wyatt, T. and Carlton, J. T. (2002). Phytoplankton introductions in European coastal waters: why are so few invasions reported? In F. Briand (Ed.). *Alien Marine Organisms Introduced by Ships in the Mediterranean and Black Seas.* Conseil International pour l'Exploration Scientifique de la Mer Méditerranée, *Workshop Monographs, 20,* 41-46.

Xiao, X., Sogge, H., Lagesen, K., Tooming-Klunderud, A., Jakobsen, K. S. and Rohrlack, T. (2014). Use of High Throughput Sequencing and Light Microscopy Show Contrasting Results in a Study of Phytoplankton Occurrence in a Freshwater Environment. *PloS one*, *9*, e106510.

Yoshida, M., Fukuyo, Y., Murase, T., and Ikegami, T. (1996). On-board observations of phytoplankton viability in ships' ballast tanks under critical light and temperature conditions. In T. Yasumoto, Y. Oshima and

Y. Fukuyo (Eds.), *Harmful and Toxic Algal Blooms* (pp. 205-208). Intergovernmental Oceanographic Commission of UNESCO.

In: Biological Invasions
Editor: Rebecca Waterman
ISBN: 978-1-63482-019-6

Chapter 2

BIOLOGICAL INVASIONS OF MALARIA PARASITES AND THEIR BIRD HOSTS

Luz García-Longoria, Sergio Magallanes, Florentino de Lope and Alfonso Marzal*
Department of Anatomy, Cellular Biology and Zoology,
University of Extremadura, Spain

ABSTRACT

In last centuries, many bird species have been introduced worldwide. Some of these bird species have become successful invaders in the new areas, thus representing a serious problem for wildlife, agriculture and even human health. Parasites have been proposed to play a major role on bird invasions, facilitating or limiting colonization and spread of their avian hosts in new continents and islands. Several hypotheses could be applied to explain the role of parasites in avian invasions. The *Novel Weapon Hypotheses* (*NWH*) builds on the idea that invasive bird species gain advantages over native birds by bringing their own parasites to the new environments against which the introduced species but not the natives have evolved defenses. These co-introduced parasites may switch to native hosts and spread in the new communities, hence becoming themselves invasive parasites provoking serious damages to indigenous bird species. But the role of parasites in invasions may extend well beyond such direct effects, and hence some indirect species may also be

* Email to: amarzal@unex.es.

expected. In this sense, the *Enemy Release Hypothesis (ERH)* poses that non-native bird species could become invaders because they have lost their co-evolved malaria parasites during the process of colonization, and thus they may increase their competitive ability and displace native species in the new areas. Conversely, the *Biotic Resistance Hypothesis* states that native parasites in the indigenous species may reduce the fitness of the potential bird invader and prevent its spread and establishment.

Malaria species are among the most pathogenic species of poultry and wildlife birds, being responsible of economic losses, mass mortality, population declines and even extinctions of many bird species worldwide after its introduction outside its native range. For all these reasons, the International Union for Conservation of Nature classifies avian *Plasmodium* to be among the 100 of the world's worst invasive alien species. In this review we will recapitulate the main scientific contributions showing the role of avian malaria parasites in the global spread of their bird hosts, evaluating the characteristics of both bird host and malaria parasite to become a fruitful invader. We will also highlight the extinctions and endangerment of numerous native species provoked by avian malaria invasions through the world (Hawaii, New Zealand, Galápagos Islands), and new menaces of avian malaria invasions in America.

INTRODUCTION

Despite the economic and ecological importance of invasive species and the efforts from scientists to understand biological invasions, the mechanisms that allow one species to become invasive are still poorly understood. Many introduced species fail to establish or to spread significantly, but many others become successful colonizers; hence, we wonder what make a species a successful invader. Parasites and other pathogens have been proposed to play an important role on the invasive process, facilitating or limiting colonization and spread of their hosts in new continents and islands (Tompkins et al. 2011). As a consequence, biological invasions of parasites may results in changes in community structure and biodiversity, with introduced pathogens disrupting native communities via both direct and indirect effects (White et al. 2006).

Much research has focused on the direct impact that parasites have on biological invasion (Hatcher et al. 2012). The *Novel Weapon Hypotheses* (*NWH*) builds on the idea that invasive species bring their own parasites to the new environments against which the introduced species but not the natives have evolved defenses (Callaway and Ridenour 2004; Prenter et al. 2004).

Consequently, individuals and populations in their native areas can be seriously damaged by the pathogens brought with invaders. One important issue in this direct effect is that co-introduced parasites, those which have been transported with an alien host to a new locality, may switch to native hosts and spread in the new communities, hence becoming themselves invasive parasites. Novel weapons have played a role in the history of the spread of humans around the world, with many examples of colonists bringing diseases that become epidemic in susceptible native populations (Diamond 1997). As well as other endemic European diseases such as measles, influenza, typhus and bubonic plague, smallpox played a decisive role in European conquests in the Americas. These novel weapons spread quickly, from tribe to tribe, killing an estimated 95% of the indigenous population far in advance of the European themselves (Diamond 1997).

But the role of parasites in invasions may extend well beyond such direct effects. As parasites are involved in interactions at all trophic levels (Kuris et al. 2008; Hatcher et al. 2012), indirect effects on species may also be expected (Hatcher et al. 2012). In this sense, the *Enemy Release Hypothesis (ERH)* poses that non-native species become successfully established because they are freed from their co-evolved pathogens and parasites (Tourchin et al. 2002; Colautti et al. 2004). Because parasites can exert detrimental effects on their hosts and reduce host population abundance, density and spread (Anderson and May 1992), a decrease in parasite pressure could allow introduced species to decrease their investment in defenses and maximize their capital in growth and reproduction, thus increasing their competitive ability and displace native species (Bloosey and Nötzold 1995; Lee and Klasing 2004). The reverse of the previous hypothesis is the *Biotic Resistance Hypothesis* (Levine et al. 2004), occurring when native parasites in the indigenous species may reduce the fitness of the potential bird invader and prevent its spread and establishment (Ricklefs 2010).

In last centuries, several bird species have been successfully introduced and become invaders in many parts of the world, both naturally and with human assistance (Blackburn et al. 2009). These invasions may represent a serious problem for wildlife, agriculture and even human health. Despite the large knowledge on the distribution, ecology, and evolution of birds could provide excellent opportunities for examining the factors that facilitate introductions, their biological invasions have received relatively little attention in compare to other organisms. Just recently some researchers have focused on avian species addressing important questions about the factors that facilitate invasions. In this line, some malaria and related haemosporidian parasites (mainly genera *Plasmodium* and *Haemoproteus*) have been proposed to play a major role on the successful colonization of exotic bird species. These parasites show a complex life cycle (Valkiūnas 2005), where the presence of a vector (biting midges, louse flies and a large number of mosquito species) is needed for transmitting the disease (Valkiūnas 2005; Martinsen et al. 2008). Haemosporidian parasites are widespread and harmful organisms (Valkiūnas 2005), provoking detrimental effects on the life-history of their avian hosts by reducing survival (Dawson and Bortolotti 2000; Valkiūnas 2005; Marzal et al. 2008), reproductive success (Merino et al. 2000; MacDougall-Shackleton et al. 2002; Marzal et al. 2005; Tomás et al. 2007), and body condition (Valkiūnas et al. 2006; Palinauskas et al. 2008). Its broad distribution, its complex cycle life and its lethal effects make malaria parasites one of the most dangerous blood parasites in nature with no preventive treatment. For all these reasons, the International Union for Conservation of Nature (IUCN) classifies some *Plasmodium* species within the 100 world's worst invasive species (Lowe et al. 2000).

Biological Invasions of Malaria Parasites and Their Bird Hosts

More than 1,400 human attempts to introduce 400 bird species have been recorded worldwide. Most of these trials came along with European colonizations in the 18th and 19th centuries, when bird species originated from temperate regions were introduced by settlers to various countries (Sodhi 2010).

Introduced birds may become overabundant posing a threat to agriculture, the environment or human health and safety. Some species such as sparrows and crows can damage crops. In addition, other invasive bird species may spread disease to humans. For example, it is thought that the West Nile virus, a pathogen mainly infecting birds but also causing mortality to humans, was introduced to New York by migratory or invasive bird species (Calisher 2000). Another illustration of zoonosis linked to birds is the bird flu virus (H5N1), which has been registered transported by invasive bird species (Webster and Govorkova 2006).

But not all of the bird introductions have resulted in established populations of invasive birds in the new regions. In consequence, some life history traits and ecological attributes, such as colonial nesting, behavioral flexibility and feeding preferences (e.g. granivorous), could allow some alien species to maintain high survival and reproductive success in new locations and to become successful invaders (Sodhi 2010). In addition to these characteristics, parasites have been proposed to be a key component in biological invasions. Hence, parasitic infections (or sometimes, their absence) may facilitate or limit invasions impacting native species via both direct and indirect effects (Hatcher et al. 2012).

The above mentioned hypotheses *(Novel Weapon Hypothesis, Enemy Release Hypothesis* and *Biotic Resistance Hypothesis)* have been proposed to explain the role of parasites in invasions in bird – parasite systems. Next we present a review of the latest scientific contributions explaining the importance of malaria parasites in avian invasions, and some of the most dramatic examples of the outcomes of the introduction of malaria parasites in bird communities.

Avian Malaria Parasites as Emerging Infectious Diseases: The Novel Weapon Hypothesis

Introduction of Avian Malaria in Hawaiian Islands

Hawaiian Archipelago comprises hundreds of islands located in the north part of Polynesia in the Pacific Ocean and are distributed along 2400 km. Hawaiian Islands represent an important source of biodiversity. Because Hawaii owns eleven out thirteen climatic areas present in the world, its fauna is of an ecological and evolutionary significance, with a considerable high number of endemic species (Carlquist 1980). Although avian malaria parasites were likely to have been present in migratory birds that visit Hawaii annually, Hawaiian bird communities were free from this disease because the absence of vectors transmitting the disease. However, since its discovery by Captain Cook in 1774 (Obeyesekere 1994) over two-thirds of the endemic bird species have become extinct. The expansion of the commercial sector to different countries like USA during the 19th century provoked an increase of movements of ships along the entire Hawaiian coast. Independently to the economic benefits for Hawaii, these enlarge of movements among islands and continents caused an increasing of opportunities for spread by parasites and pathogens. As a consequence, the mosquito *Culex quinquefasciatus*, primary vector of avian malaria, was introduced to the Hawaiian islands in 1826, presumably in water casks from shipping vessel HMS Wellington recently sailed from Mexico (LaPointe et al. 2005; Beadell et al. 2006). In that year, the crew of the ship dumped the mosquito-laden remnants of their water kegs into a Hawaiian stream to refill them with fresh water (Leighton 2002). Hence, mosquito populations established and spread in Hawaii, representing the first and one of the most serious insect pests arriving to Hawaii.

The question remaining is where do the parasites come from? It has been proposed that reservoirs of avian malaria parasite started with the co-transported of exotic host species during colonial times. Additionally, it has been suggested that thousands of ducks and shorebirds that annually migrate to Hawaii from the Arctic may act as natural reservoirs (Beadell et al. 2006). Many endemic Hawaiian species are highly susceptible to avian malaria (Atkinson et al. 2005). In contrast, malaria has minimal impact on nonnative birds, and these nonnative species may act as effective reservoirs for avian malaria (Atkinson et al. 2005, Atkinson and Samuel 2010). Wherever they come from, it is clear that the introduction of nonnative bird species in Hawaii had dramatic effects for native avifauna. From the moment that avian malaria and its vector were successfully established in the archipelago (around 1920s), a wave of extinctions and endangerment among Hawaiian forest birds, which began by the 1920s (Van Riper et al. 1986).

More than a half of over 100 endemic bird taxa in Hawaii have been driven extinct by a combination of habitat loss, introduced species and diseases (Olson and James 1982, LaPointe 2005). In 1941, six species of *Plasmodium* genus were recorded for the first time from endemic and exotic birds in Hawaii (see review in Laird and Van Riper 1981). Since then, the number of studies analyzing the diversity and virulence of these blood parasites in native bird species has increase significantly. For example, Van Riper et al. (1986) conducted laboratory and field experiments in 16 different areas of Hawaiian Islands to determine the impact of avian malaria on the forest birds. As expected, they found that native forest birds were more susceptible to malaria parasite than introduced species, explaining the alarming fall in the number of native species during the previous years and concluding that avian malaria was responsible for the wave of extinctions of native bird species.

Moreover, *Plasmodium* parasites may also act as a barrier to the movement of the native bird species (Dobson and Hudson 1986) provoking a reduction of the habitat in endemic bird species. *Culex* mosquitoes are abundant at 1500 m.a.s.l., where the most optimal breeding grounds of native bird species are. Therefore, many native birds can no longer breed in their historic breeding grounds at lower elevations because of avian malaria, forcing them to breed in higher sub-optimal elevations where food and cover may be scarce. This situation has provoked a population decline and restriction of the distribution of many susceptible Hawaiian bird species (Samuel et al. 2010). Nowadays this situation is worsening because it has been reported that prevalence of malaria in Hawaiian forest birds at 1900 ma.s.l. has more than double over a decade, demonstrating the rapid spread of this parasite (Freed et al. 2005).

In contrast, Kilpatrick and LaPointe (2006) found that the chronic avian malaria infection by *Plasmodium relictum* did not have any negatively effect in clutch size, hatching success, fledging mass, number of nestlings fledged, nesting success and minimum fledgling survival of the native Amakihi (*Hemignathus virens*). In fact, they found a higher nestling success and higher survival to the next year in birds which parents carried a malaria chronic infection. They also found that the reproduction and survival of infected birds were sufficient for a per-capita population growth rate >1, which suggests that chronically infected Hawaii Amakihi could support a growing population.

Summarizing, avian malaria *Plasmodium relictum* has been largely responsible for the recent wave of extinctions and endangerment among Hawaiian forest bird. This is the best-known example of the catastrophic decline in endemic avifauna provoked by avian malaria invasion. The introduction of non-native bird species co-transporting avian malaria, and also acting as reservoirs for the parasite, as well as the colonization of its mosquito vector have provoked the decimation of many native Hawaiian birds. Future research should be directed at investigations controlling this avian disease in native environments. In this sense, reducing artificial larval mosquito habitat and eliminating feral pigs could effectively reduce mosquito populations and hence limit the transmission of this disease to native birds. This avian disease management must be coordinated over broad landscapes and will necessitate the cooperation of researchers, communities, and governmental agencies.

Avian Malaria Outbreaks in New Zealand

New Zealand is an island country situated 1,500 kilometers east of Australia across the Tasman Sea and 1,000 kilometers south of the Pacific island areas of New Caledonia, Fiji, and Tonga. Because of its remoteness, it was one of the last lands to be settled by humans (Higham et al. 1999). Due to this historic and geographical isolation, New Zealand has highly endemic fauna (Daugherty et al. 1993). Since its human colonization in about 1280-1300 AD (Wilmshurst et al. 2008), many different exotic species have invaded New Zealand. This problem became worse during the 20th century, with the exponential rise of ships and aircrafts moving around the world, thus increasing the opportunities for spread of parasites and pathogens. This has resulted in a serious threat regarding the possibility of avian malaria outbreak in New Zealand involving novel parasites (Tompkins and Gleeson 2006)

The first records of avian malaria in New Zealand date from the earliest years of 20th century, when it was predominant detected in imported bird species (Doré 1920). Some years later, Laird (1950) examined blood smears from 50 different bird species, including some introduced passerines. Although he described blood stages of haemosporidian infections in introduced birds, he was not able to find any evidence of malaria infection in endemic birds. However, he postulated the possibility of introduced avian malaria as a potential factor in local native bird extinctions, and he suggested wider studies on bird haematozoa because the low sample size of his study. In 1970s, Fallis et al. (1976) scanned thin blood smears from 43 native and introduced bird species, but he was unable to detect avian malaria in the examined birds, most probably because the low sample size of 1-10 per bird species.

In contrast to these initial studies, recent investigations have detected *Plasmodium* spp. in New Zealand birds, suggesting that avian malaria could be an emerging threat to New Zealand avifauna (see reviews in Barraclough et al. 2012; Schoener et al. 2014). For example, an outbreak of malaria among endemic Yellowheads (*Mohoua ochrocepahala*) held in Christchurch Wildlife Park killed more than 90% of the population (Derraik 2006). Also, infection with *Plasmodium* spp. provoked the death of five Yellowheads at Orana Wildlife Park (Alleya et al. 2008). More recently, Howe et al. (2012) examined three introduced, two native and seven endemic avian species for the presence of *Plasmodium* spp., revealing six different *Plasmodium* lineages infecting New Zealand bird species. In addition, they reported the death of native and exotic bird species due to acute *Plasmodium* spp. infection. Also, Ewen et al. (2012) detected eight *Plasmodium* parasites infecting six native and four exotic bird species. Likewise, Baillie et al. (2012) have documented four *Plasmodium* species infecting the endemic New Zealand passerine Bellbird *Anthornis melanura*, which three of them were exotic parasite lineages (*Plasmodium* (*Huffia*) *elongatum*, *Plasmodium* (*Novyella*) *vaughani* and *Plasmodium* (*Haemamoeba*) *relictum*). As an overall, these and other avian malaria parasites have been found in 35 different endemic, native and exotic bird species in New Zealand, and they have been diagnosed as causing the death in threatened species (Schoener et al. 2014), thus constituting a serious impact to native avifauna.

Most of these avian malaria parasites found in New Zealand are not closely related to one another, and are not known to be endemic to New Zealand (Ewen et al. 2012; Schoener et al. 2014). Therefore, it has been suggested that these parasites were co-transported with their exotic host species during colonial times (Thomson 1922) and arrived in New Zealand in multiple independent events (Ewen et al. 2012). However, not all the introduced species become fruitful invaders. Hence, some features of parasites may determine whether or not they are successfully co-established into new regions. For example, it has been proposed that fruitful invasive parasites are predicted to be generalist, because they are more likely to encounter new hosts that they can infect. Also, they are expected to have relatively high prevalence and low virulence in the natural host, because this would mean higher probabilities of being present in host founders for release, but not provoking host mortality during the invasion process. And finally, competent vectors for these parasites should be present in the new areas to ensure malaria transmission. Following this idea, Ewen et al. (2012) recently tested the traits of successful established avian malaria parasites co-introduced along with numerous host species in New Zealand. They showed that exotic parasites that successfully invaded this remote island archipelago were more globally generalist (were recorded from most host species and were also more widely distributed outside New Zealand) than other known parasites not found in New Zealand. They also found that avian malaria parasites introduced to New Zealand were more prevalent in their native range in Europe than malaria parasites found infecting birds in Europe but not in New Zealand. Finally, it is known that 12 indigenous and four exotic species of mosquitoes are established in New Zealand (Derraik 2004a), being all four alien species potential disease vectors. *Culex quinquefasciatus*, the primary vector of avian malaria in Hawaii, is among them (Laird 1995). This exotic mosquito has spread rapidly in New Zealand (Tompkins and Gleeson 2006), and is likely to be the vector responsible for some avian malaria outbreaks in New Zealand (Derraik 2004b). The entrance pathway of the exotic mosquito species is anthropogenic, mainly by aircrafts and ships coming from Asia and South Pacific (Derraik 2004a). Therefore, all these features seem to have facilitated the invasion of some avian malaria parasites in New Zealand.

To recapitulate, in the last decade avian malaria has aroused increasing interest as an emerging infectious disease in New Zealand. Introduced birds carrying *Plasmodium* spp. and the spread of exotic mosquitoes now represent a serious menace because it can boost the parasitemia in infected species, increase its host range in areas where *Plasmodium* is already present, and/or increase its geographical range in New Zealand (Tompkins and Poulin 2006). It is necessary, therefore, the establishment of exhaustive exotic and endemic species monitoring and border surveillance in order to prevent new avian malaria outbreaks in New Zealand.

Studies of Avian Malaria in Galápagos Islands

Galápagos Islands are an archipelago of volcanic islands located approximately 1000 km west of continental Ecuador, of which they are a part. Because they are remote located and were absent of permanent human settlements until 19th century, much of the original biodiversity remains intact, with only 5% species loss (Gibbs et al. 1999). These islands are of remarkable ecological and evolutionary significance. Their unique sets of environmental conditions and the ecological isolation have facilitated the development of high levels of endemism and some of the most unique life forms on the planet. In 1835 a young Charles Darwin was inspired on the islands when collected material that underpinned On the Origin of Species. But this historic visit, in addition to help us in our current understanding of ecology and evolution, also caused this Archipelago began to receive international recognition. Since then, the native ecosystems on the Galápagos Islands are suffering severe destruction as a result of the introduction of exotic species, brought accidentally or willingly to the islands by humans (Mauchamp et al. 1998). Feral goats, pigs, dogs, cats, rats, and many other species have altered ecosystem characteristics and driven endemic species, such as several species of the Galápagos tortoise, to extinction (Schofield 1989; Mauchamp 1997).

As they have evolved in an isolated system, island populations are prone to suffer more from introduced pathogens than mainland populations (Fromont et al. 2001). Unlike Hawaii and other remote island archipelagos, wildlife populations from Galápagos Islands are currently concerned regarding the introduction of diseases because their isolation and high endemism. The introduction of pathogens, which may be highly infectious for the immunologically naïve native birds, could be a cause of major population declines and extinctions in Galápagos avifauna. Although the Galápagos archipelago still possesses its entire endemic avifauna, several of these bird populations are in decline. For example, the Galápagos penguin *(Spheniscus mendiculus)*, an endemic species to the Galápagos Islands, is classified as Endangered due to small population size and restricted geographical range (BirdLife International 2008). Moreover, Galápagos penguins exhibit low levels of genetic diversity (Nims et al. 2008) and very low variation in major histocompatibility complex (MHC) genes (Bollmer et al. 2007), which could contribute to the susceptibility of the population to infectious disease.

Previous studies based on microscopic and molecular screening of many species, including the Galápagos penguin, did not detect malaria parasites in Galápagos birds (Miller et al. 2001a; Travis et al. 2006). But later Levin et al. (2009) identified 19 (5%) of 362 penguins as positives for *Plasmodium*, thus being the first report of a *Plasmodium* blood parasite found in the Galápagos Archipelago. In this study, the *Plasmodium* infections were widely distributed across nine sites of five islands in the Archipelago. Moreover, the phylogenetic analyses of this new parasite suggested a close relationship to *P. elongatum*, one of the most dangerous lineages of *Plasmodium* that has been known to cause severe mortality and morbidity in captive penguin populations (Fleischman et al. 1968). However, in all these studies penguins seemed to be healthy, without any obvious impact on their survival and reproduction. Furthermore, they were not able to identify neither the competent vector for *Plasmodium* nor the sites where those infections were contracted. Also, the microscopic evaluation of blood smears of PCR-positive individuals did not show any gametocytes or other blood stages, and hence whether the parasite could complete its life cycle and form gametocytes in Galápagos penguins to ensure its transmission remained unclear. In consequence, many questions about the parasite's mode and time of arrival, pathogenicity, possible vectors, and transmission dynamics within the archipelago remained unsolved at this point.

Later, Santiago-Alarcón et al. (2010) studied the phylogenetic relationship of *Haemoproteus* parasites in the endemic Galápagos dove (*Zenaida galapagoensis*), a Columbiform that suffers from high prevalences and intensities of haemosporidian infections (Padilla et al. 2004; Santiago-Alarcón et al. 2008). They showed that haemosporidian parasites infecting endemic Galápagos doves were not unique to the archipelago and were widely distributed across the islands, suggesting multiple colonizations of parasite lineages existing on the continent.

More recently, Levin et al. (2013) screened 3726 endemic birds representing 22 species for *Plasmodium* spp. through a combination of molecular and microscopy techniques. They found four different genetic lineages of *Plasmodium* parasites infecting Galápagos birds (lineages A, B, C, D). Lineage A was first described only from Galápagos Penguins (Levin et al. 2009), while lineages B, C and D were new findings. These later lineages were each detected at one site and at one time, implying that they were transient infections of parasites not established on the archipelago and birds may be dead-end hosts for these *Plasmodium* lineages. These lineages may not have established due to very small numbers of infected individuals or to possible incompatibilities among host, parasite, and vector. In contrast to the lack of evidence for establishment of these lineages, lineage A infected more than one species at multiple locations and across multiple years, which suggested lineage A is established and is being transmitted regularly in the archipelago. However, they did not find gametocytes in blood smears of any PCR positive individuals, and hence could not confirm that the parasite is completing development and producing gametocytes. They also revealed lineage matches between *Plasmodium* spp. found in Bobolinks *(Dolichonyx oryzivorus)* with Galápagos lineages B and C, which may indicate that North American breeding Bobolinks, which regularly migrate through Galápagos, could be responsible for infecting local vectors that transmit the parasite to local birds. In this sense, the only known competent vector for *Plasmodium* parasites present in Galápagos is the mosquito *Culex quinquefasciatus*, first reported in 1989 and well established by 2003 (Whiteman et al. 2005), which it has been linked to the disappearance of resident penguins on the north shore of the island of Santa Cruz in the Galápagos Archipelago (Miller et al. 2001b). Besides this natural colonization of parasites transported by migratory birds, alternative routes linked to human activities may not be underestimated because they usually have a larger effect on probability of introductions of invasive species (Gardener and Grenier 2011).

Summarizing, *Plasmodium* blood parasites had not been detected in Galápagos birds until last decade, despite previous microscopic and molecular studies, which suggests a recent arrival of this parasite. It is possible that this parasite arrived with or shortly after the recent arrival of the introduced mosquito vector, *Culex quinquefasciatus*. Although still there are no evidences on the negative effects of avian malaria parasites on the fitness of Galápagos native avifauna, the combined effects of *Plasmodium* infection and stressful El Niño events could represent a serious threaten of endemic avifauna of Galápagos Islands in light of future climate change scenarios. In consequence, there is a great need for extensive studies in Galápagos avifauna in order to characterize the malaria invasive species, identify its vector, and develop an appropriate management strategy to avoid its spread out and keep it under control.

Invasive Avian Malaria in Native Birds of Peru

As it has been previously shown, *Plasmodium* parasites are cosmopolitan pathogens responsible for mass mortality, population declines and even extinctions of many bird species worldwide after its introduction outside its native range (Van Riper et al. 1986; Valkiūnas 2005). *Plasmodium relictum*, a widespread mosquito-transmitted agent of avian malaria, is of particular interest because of its highly virulence, genetic diversity and markedly invasive nature (LaPointe et al. 2005; Valkiūnas 2005; Beadell et al. 2006; Palinauskas et al. 2008; Marzal et al. 2011). Recent PCR-based studies have revealed a huge genetic diversity of this pathogen, showing different parasites lineages based on partial sequences of the cytochrome oxidase b gene. For example, *P. relictum* pGRW4, the parasite lineage responsible of devastating epizooties reported in Hawaii, has a broad geographical range including New Zealand, Africa, Asia and the Americas (Beadell et al. 2006; Marzal et al. 2011). It is closely related to lineage *P. relictum* pSGS1, which is a widespread and actively transmitted parasite lineage in Europe, Africa and Asia (Palinauskas et al. 2007), although recently it has also been recorded infecting native and indigenous birds in Oceania (Howe et al. 2012). Both parasites lineages might easily switch to new hosts as they spread into new areas (Beadell et al. 2006; Hellgren et al. 2009). *P. relictum* lineage pSGS1 has been reported in over 60 species of birds, but until very recently, this invasive lineage had not been reported in the mainland Americas (Beadell et al. 2006; Durrant et al. 2006; Merino et al. 2008; Marzal et al. 2011; Lacorte et al. 2013).

Marzal et al. (2014) have recently analyzed the presence of pSGS1 in Neotropical birds from two different areas of Peru, Lima and Huanuco. They examined 102 blood samples from 18 native bird species for haemosporidian infection. They found five *Haemoproteus* lineages and five *Plasmodium* lineages infecting 12 different bird species. Invasive *Plasmodium relictum* SGS1 was the most generalist and the most prevalent parasite lineage found in the study, infecting 13 individuals from eight host species (39.40 % of the total infections). Moreover, pSGS1 was also a geographically widespread parasite, being the only *Plasmodium* lineage infecting birds in both study areas. It is remarkably that this parasite lineage had not been previously reported in any of the 39 studies analyzing avian malaria and related haemosporidian parasites from 213 native and introduced bird species from 17 orders covering almost all the geographical range from North and South America (MalAvi database, version 2.0.5 June 2013, Bensch et al. 2009). Consequently, this study constituted the first report of this invasive pathogen in the mainland Americas.

This exotic parasites could cause extreme morbidity and mortality in new hosts because the lack of an evolved immune response against the indigenous parasite (Schmid-Hempel 2011). Therefore, the presence of this invasive parasite in birds of South America may represent a serious risk to this avifauna. Moreover, the high prevalence of this exotic *Plasmodium* lineage in the limit with Peruvian Amazon should warn us about the potential threat to over one-third of all bird species in the world.

Avian Malaria Parasites and the Enemy Release Hypothesis

Previous studies represent some of the most well known examples of bird invasions and biodiversity losses as a consequence of the emerging infectious diseases provoked by the invasion of malaria parasites coming along with their bird hosts. But the opposite scenario is also possible in invasions of exotic birds. The success of introduced birds when arriving in a non-native region could also be explained if parasites "missed the boat" on the trip to new areas (Paterson and Gray 1997). The *Enemy Release Hypothesis (ERH)* suggests that a flourishing colonizer is successful because it has left behind its natural predators and parasites, which usually reduced its fitness. In consequence, successful colonizers do not have to invest in anti-parasite or anti-predators defenses and can reallocate resources into other important biological functions (e.g. reproduction and growth).

Because avian malaria parasites exert negative effects on their host populations by reducing host fitness (Merino et al. 2000; Valkiūnas 2005; Marzal et al. 2005), these haemosporidian parasites provide an excellent opportunity to test the ERH. Beadell et al. (2006) were the first to study the avian blood parasites as a model to test this evolutionary hypothesis. They evaluated the prevalence and distribution of *Plasmodium* and *Haemoproteus* parasites in native and six introduced populations of the common myna *Acridotheres tristis*, a common passerine native to southern Asia that has been introduced to many parts of the world. Their findings showed that not all comparisons of introduced populations to the native populations were consistent with the predicted *Enemy Release Hypothesis*. Moreover, they found some evidence that introduced populations may have become infected with novel parasite lineages, and also that common mynas carried parasite lineages from native to introduced locations. These mixed results could be explained because the difficulty to differentiate between parasites that are native and introduced, since haemosporidian lineages infecting common mynas often did not show regional or host specificity.

Hutchings (2009) investigated if the poor parasite diversity found in the endemic Seychelles warbler (*Acrocephalus sechellensis*) in the islands of the

Indian Ocean could be due to the loss of parasites during historical colorizations and contemporary conservation driven translocations. He analyzed the prevalence and identity of haemosporidian parasites on three populations of Seychelles warbler that have been established by conservation driven translocations from the single remaining population on Cousin Island. He found evidence of enemy release, because no malaria infection was seen in two of the populations. However, a third translocated population still contained malaria. He also observed a decrease in malaria prevalence over a three-year period, and none of the birds born on the island had malaria infection. These results support a hypothesis of previous enemy release in the Seychelles warbler, leading to loss of parasite diversity.

Lima et al. (2010) also tested the ERH by analyzing the prevalence and intensity of *Plasmodium* spp. and *Haemoproteus* spp. in introduced house sparrows (*Passer domesticus*) and native urban birds of central Brazil, as well as comparison with European house sparrows. In accordance with ERH, they found that native birds from Brazil presented significantly higher parasite prevalence. They also found that house sparrows native to Europe exhibited significantly higher parasite prevalence than introduced house sparrows from Brazil, suggesting that sparrows from Brazil might have experienced a parasitic release during the process of introduction.

Similarly, Marzal et al. (2011) examined the ERH and NWH to assess the role of avian malaria parasites in the global spread of a common invasive bird, the house sparrow. They explored whether this invasive species carried its parasites to new areas, and/or acquired new local parasites. By analyzing prevalence and genetic diversity of malaria parasites in 58 localities on six continents, they showed that house sparrows did not retain their native parasites in newly colonized regions, supporting the hypothesis that the release from these natural enemies facilitated its spread in the last two centuries.

In another study, Baillie et al. (2012) documented patterns in avian malaria, *Plasmodium* spp., infecting an endemic New Zealand passerine, the bellbird *Anthornis melanura*, at multiple-host subpopulations simultaneously. They tested the ERH after a recent colonization event of the bellbird host. According to Phillips et al. (2010), they predicted that low-density founder populations either should have lower parasite species diversity and prevalence than the source population. They observed a decrease in parasite species diversity and an increase in body condition from source to founder sites, but statistical differences in the direct relationship between body condition and malaria prevalence between source and colony were weak and significant only during winter. Moreover, they did not found any significant decrease in the prevalence of *Plasmodium* parasites from source to founder populations. Hence, their findings did not provide strong support for the ERH.

More recently, Coon and Martin (2014) examined the changes in haemosporidian prevalence in house sparrows over small spatiotemporal scales to assess whether NWH or ERH facilitated the introduction success of sparrows in Kenya. They found little evidence of enemy release or use of novel weapons by range-expanding house sparrows in Kenya, suggesting that host qualities, such as sparrow's predisposition towards minimal behavioral and physiological response to parasites, may be more important to house sparrow range expansion success than parasitological phenomena such as NWH or ERH.

Avian Malaria Parasites and the Biotic Resistance Hypothesis

Conversely to the previous hypotheses, the *Biotic Resistance Hypothesis* states that a native community may have the ability to keep out invasive species. Most of studies of biotic resistance in animal communities have focused on competition and predation as mechanisms of resistance. For example, it has been shown that a native predator, the blue crab *Callinectes sapidus*, provides biotic resistance to invasion and prevents the spread and establishment of the introduced European green crab *Carcinus maenas* in eastern North America (DeRivera et al. 2005).

Additionally, parasites also have the ability to prevent animal invasions. In this sense, the presence of parasites native to the novel habitat can infect the introduced species and limit the establishment and spread of these non-native species (Levine et al. 2004). But surprisingly, despite their ubiquity and their essential role in the dynamics of invasions, the importance of parasites to create biotic resistance has been almost totally ignored and the number of studies is still very limited. For instance, MacColl and Chapman (2010) experimentally tested the contribution that parasites could make to selection against migrants in the adaptive radiation of three-spined sticklebacks (*Gasterosteus aculeatus*), showing that sticklebacks contracted higher burdens of novel parasites when introduced to novel environment, and suffered a growth cost as a direct result.

In the case of bird host – malaria parasite systems, some invasions may have failed because of native parasites that reduce the fitness of the invader and prevent its establishment. However, failed invasions are often missed in scientific studies, and so this plausible process has not been well documented.

The best example showing that avian malaria parasites could provide biotic resistance avoiding the establishment of an exotic bird species has been reviewed by Ricklefs (2010) while studying host–pathogen coevolution, secondary sympatry and species diversification in birds from the West Indies. Speciation process requires the achieving of reproductive isolation followed by secondary sympatry. It has been suggested that secondary sympatry might be prevented by apparent competition mediated through pathogens that are locally coevolved with one population of host and are pathogenic in sister populations (Ricklefs and Bermingham 2007; Ricklefs 2010). For example, it has been shown that *Plasmodium* and *Haemoproteus* parasites exhibit significant avian host species x island interactions in prevalence in the Lesser Antilles (Apanius et al. 2000; Fallon et al. 2003). This findings suggests that these parasites have evolved locally to host populations, and hence local parasites could prevent the invasion of an islands of host sister populations coming from other islands in the archipelago.

We currently lack of more direct test of the hypothesis that avian malaria parasites can prevent introductions of alien species. Future studies doing reciprocal translocations of both avian malaria parasites and their hosts conducting cross-infections under controlled conditions in laboratories would be very informative with respect to testing the *biotic resistance hypothesis* in this host – parasite system.

CONCLUSION AND FUTURE DIRECTIONS

We are living in times of unprecedented global changes. Although species invasions have periodically occurred in the history of life, human-assisted invasions have increased enormously in last decades. In combination to the introduction of nonnative species, other anthropogenic alterations such as habitat destruction or fragmentation have provoked changes in host-parasite ecology, resulting in outbreaks of human, domestic animal and wildlife Emerging Infectious Diseases. Detrimental effects on native communities usually accompany human introduction of species to new geographical regions. Here we have shown different examples of native bird populations that have been decimated by introduced mosquito transmitted avian malaria. The emergence of this and other wildlife diseases may further increase the zoonotic threat to human health by amplifying transmission rates of zoonotic pathogens.

Our current knowledge relevant to the disease effects of biological invasions is still limited, and many critical questions, such as the responses of the most devastating invasive species to relevant components of global change, have yet to be addressed. Further investigation and exploration are needed in order to gain a better understanding and control of these biological invasions and design interventions. These investigations require an integrated, multidisciplinary approach. Future cooperation among wildlife ecologists, conservation biologists, environmental biologists, veterinarians and medical scientists will provide a fruitful knowledge for developing comprehensive risk assessment to aid governments and decision makers. In addition, strict border control biosecurity regimes and controlled human traffic to remote populations or endangered species become essential to prevent accidental or deliberate introductions.

Acknowledgments

The authors thank NOVAPUBLISHER for the invitation to write this review. The authors also are grateful to J. Balbontín and J.L. Pérez-Bote for helpful suggestions regarding the manuscript. Spanish Ministry of Economy and Competitiveness (CGL2012-36665, BES-2010-030295, BES-2013-063589) provided funding for this research.

References

Alleya M, Fairley R, Martinc D, Howea L, Atkinsond T (2008) An outbreak of avian malaria in captive yellowheads/mohua (*Mohoua ochrocephala*). *New Zeal Vet J* 56:247–251.

Anderson RM, May RM (1992) *Indirectly transmitted microparasites. In Infectious Diseases of Humans: Dynamics and Control.* Oxford Scientific, Oxford, UK.

Atkinson C, Lease J, Dusek R, Samuel MD (2005) Prevalence of pox-like lesions and malaria in forest bird communities on leeward Mauna Loa Volcano, Hawaii. *The Condor* 107:537–546.

Atkinson C, Samuel MD (2010) Avian malaria *Plasmodium relictum* in native Hawaiian forest birds: epizootiology and demographic impacts on Apapane *Himatione sanguinea*. *J Avian Biol* 41:357–366.

Apanius V, Yorinks N, Bermingham E, Ricklefs RE (2000) Island and taxon effects in parasitism and resistance of Lesser Antillean birds. *Ecology* 81:1959–1969.

Baillie SM, Gudex-Cross D, Barraclough RK, Blanchard W, Brunton DH (2012) Patterns in avian malaria at founder and source populations of an endemic New Zealand passerine. *Parasitol Research* 111:2077–2089.

Barraclough TG, Balbi KJ, Ellis RJ (2012) Evolving Concepts of Bacterial Species. *Evol Biol* 39:148–157.

Beadell JS, Ishtiaq F, Baker AJ, et al. (2006) Prevalence and evolutionary relationships of haematozoan parasites in native versus introduced populations of common myna *Acridotheres tristis*. *Proc Biol Sci* 273:587–594.

Bensch S, Hellgren O, Pérez-Tris J (2009) MalAvi: a public database of malaria parasites and related haemosporidians in avian hosts based on mitochondrial cytochrome b lineages. *Mol Ecol Resour* 9:1353–1358.

BirdLife International (2008) *Critically Endangered birds: a global audit, BirdLife I.* Cambridge, UK.

Blackburn TM, Lockwood JL, Cassey P (2009) *Avian Invasions: The Ecology and Evolution of Exotic Birds.* Oxford Avian Biol Ser, Oxford University Press, UK.

Blossey B, Nötzold (1995) Evolution of increased competitive ability in invasive non-indigenous plants: a hypothesis. *J Ecol* 83:887-889.

Bollmer JL, Vargas FH, Parker PG (2007) Low MHC variation in the endangered Galápagos penguin (*Spheniscus mendiculus*). *Immunogenetics* 59:593–602.

Calisher CHC (2000) West Nile Virus in the New World: Appearance, Persistence, and Adaptation to a New Econiche—An Opportunity Taken. *Viral Immunol* 13:411–414.

Callaway RM, Ridenour WM (2004) Novel weapons: invasive success and the evolution of increased competitive ability. *Front Ecol Environ* 2:436–443.

Carlquist S (1980) *Hawaii: a natural history. Geology, climate, native flora and fauna above the shoreline.* Honolulu, SB Printers.

Colautti RI, Ricciardi A, Grigorovich IA, MacIsaac HJ (2004) Is invasion success explained by the enemy release hypothesis? *Ecol Lett* 7:721–733.

Coon C, Martin L (2014) Patterns of haemosporidian prevalence along a range expansion in introduced Kenyan House Sparrow (*Passer domesticus*). *J Avian Biol* 45:34–42.

Daugherty JR, Rai R, El Berry HM, Cooper TG (1993) Regulatory circuit for responses of nitrogen catabolic gene expression to the GLN3 and DAL80 proteins and nitrogen catabolite repression in *Saccharomyces cerevisiae*. *J Bacteriol* 175:64–73.

Dawson RD, Bortolotti GR (2000) Effects of hematozoan parasites on condition and return rates of American kestrels. *Auk* 117:373–380.

DeRivera CE, Ruiz GM, Hines AH, Jivoff P (2005) Biotic resistance to invasion: Native predator limits abundance and distribution of an introduced crab. *Ecology* 86:3364–3376.

Derraik JGB (2004a) A survey of the mosquito (Diptera: Culicidae) fauna of the Auckland Zoological park. *New Zeal Entomol* 27:51–55.

Derraik JGB (2004b) Exotic mosquitoes in New Zealand: A review of species intercepted their pathways and ports of entry. *Aust N Z J Public Health* 28:433–444.

Derraik JGB (2006) Bitten birds. Piecing together the avian malaria puzzle. *BNZ* 65:16–17.

Diamond J (1997) *Guns, germs, and steel: the fates of human societies.* W W Norton, New York, USA.

Dobson AP, Hudson PJ (1986) Parasites, disease and the structure of ecological communities. *Trends Ecol Evol* 1:11–15.

Doré AB (1920) The occurrence of malaria in the native ground lark. Journal of Science and Technology. *New Zeal J Sci Technol* 3:118–119.

Durrant KL, Beadell JS, Ishtiaq F, et al. (2006) Avian hematozoa in South America: a comparasion of temperate and tropical zones. *Ornithol Monogr* 60:98-102.

Ewen JG, Bensch S, Blackburn TM, Bonneaud CB, Ruth C, Phillip C, Rohan H, Pérez-Tris J (2012) Establishment of exotic parasites: The origins and characteristics of an avian malaria community in an isolated island avifauna. *Ecol Lett* 15:1112–1119.

Fallis A, Bisset S, Allison F (1976) *Leucocytozoon tawaki* sp (Eucoccidia: Leucocytozoidae) from the penguin *Eudyptes pachyrynchus*, and preliminary observations on its development in *Austrosimulium* spp. (Diptera: Dimuliidae). *New Zeal J Zool* 3:11–16.

Fallon SM, Bermingham E, Ricklefs RE (2003) Island and taxon effects in parasitism revisited: Avian malaria in the Lesser Antilles. *Evolution* 57:606–615.

Fleischman RW, Squire RA, Sladen WJL, Melby EC (1968) Malaria (*Plasmodium elongatum*) in captive African penguins (*Spheniscus demersus*). *J Am Vet Med Assoc* 153:928–935.

Freed LA, Cann RL, Goff ML, Kuntz WA, Bodner GR (2005) Increase in avian malaria at upper elevation in Hawai'i. *Condor* 107:753–764.

Fromont E, Morvilliers L, Artois M, Pontier D (2001) Parasite richness and abundance in insular and mainland feral cats: insularity or density? *Parasitology* 123:143–151.

Gardener M, Grenier C (2011) Linking livelihoods and conservation: challenges facing the Galápagos Islands. In: Baldacchino B, Niles D (eds) *Island futures and development across the Asia-Pacific region,* Global Environmental Studies, Springer, pp 73–85

Gibbs JN, Lipscombe MA, Peace JA (1999) The impact of *Phytophthora* disease on riparian populations of common alder (*Alnus glutinosa*) in southern Britain. *Eur J For Pathol* 29:39–50.

Hatcher MJ, Dick JTA, Dunn AM (2012) Diverse effects of parasites in ecosystems: Linking interdependent processes. *Front Ecol Environ* 10:186–194.

Hellgren O, Pérez-Tris J, Bensch S (2009) A jack-of-all-trades and still a master of some: prevalence and host range in avian malaria and related blood parasites. *Ecology* 90:2840–2849.

Higham T, Anderson A, Jacomb C (1999) Dating the first New Zealanders : The chronology of Wairau Bar. *Antiquity* 73:420-426.

Howe L, Castro IC, Schoener ER, Hunter SB, Rosemary K, Alley MR (2012) Malaria parasites (*Plasmodium* spp.) infecting introduced, native and endemic New Zealand birds. *Parasitol Res* 110:913–923.

Hutchings K (2009) *Parasite-mediated selection in an island endemic, the Seychelles warbler (Acrocephalus sechellensis).* PhD dissertation, University of East Anglia, Norwich, UK.

Kilpatrick AM, LaPointe DA (2006) Effects of chronic avian malaria (*Plasmodium relictum*) infection on reproductive success of Hawaii Amakihi (*Hemignathus virens*). *Auk* 123:764–774.

Kuris AM, Hechinger RF, Shaw JC, et al. (2008) Ecosystem energetic implications of parasite and free-living biomass in three estuaries. *Nature* 454:515–518.

Lacorte GA, Flix GMF, Pinheiro RRB, et al. (2013) Exploring the Diversity and Distribution of Neotropical Avian Malaria Parasites - A Molecular Survey from Southeast Brazil. *PLoS One* 3:7-10.

Laird M (1950) Some blood parasites of New Zealand birds. *Publ Zool from Victoria Univ Coll* 5:1–20.

Laird M (1995) Background and findings of the 1993-94 New Zealand mosquito survey. *New Zeal Entomol* 18:77–90.

Laird M, Van Riper C (1981) Questionable reports of *Plasmodium* from birds in Hawaii, with the recognition of *P. relictum* ssp. *capistranoae* (Russell, 1932) as the avian malaria parasite there. *Appl Anim Behav Sci* 66:159–165.

LaPointe DA, Goff ML, Atkinson CT (2005) Comparative susceptibility of introduced forest-dwelling mosquitoes in Hawai'i to avian malaria, *Plasmodium relictum*. *J Parasitol* 91:843–849.

Lee KA, Klasing KC (2004) A role for immunology in invasion biology. *Trends Ecol Evol* 19:523–529.

Leighton FA (2002) Pathogens and Disease. In Hoffman et al. (eds) *Handbook of Ecotoxicology,* CRC Press, Boca Ratón, USA.

Levin II, Outlaw DC, Vargas FH, Parker PG (2009) *Plasmodium* blood parasite found in endangered Galápagos penguins (*Spheniscus mendiculus*). *Biol Conserv* 142:3191–3195.

Levin II, Zwiers P, Deem S. L, et al. (2013) Multiple lineages of Avian malaria parasites (*Plasmodium*) in the Galápagos Islands and evidence for arrival via migratory birds. *Conserv Biol* 27:1366–1377.

Levine JM, Adler PB, Yelenik SG (2004) A meta-analysis of biotic resistance to exotic plant invasions. *Ecol Lett* 7:975–989.

Lima M, Simpson L, Fecchio A, Kyaw C (2010) Low prevalence of haemosporidian parasites in the introduced house sparrow (*Passer domesticus*) in Brazil. *Acta Parasitol* 55:297–303.

Lowe S, Browne M, Boudjelas S, De Poorter M (2000) 100 of the World's Worst Invasive Alien Species. *A selection from the Global Invasive Species Database.* IUCN/SSC Invasive Species Specialist Group (ISSG).

MacColl ADC, Chapman SM (2010) Parasites can cause selection against migrants following dispersal between environments. *Funct Ecol* 24:847–856.

MacDougall-Shackleton EA, Derryberry EP, Hahn TP (2002) Nonlocal male mountain white-crowned sparrows have lower paternity and higher parasite loads than males singing local dialect. *Behav Ecol* 13:682–689.

Martinsen ES, Perkins SL, Schall JJ (2008) A three-genome phylogeny of malaria parasites (*Plasmodium* and closely related genera): evolution of life-history traits and host switches. *Mol Phylogenet Evol* 47:261–273.

Marzal A, de Lope F, Navarro C, Møller AP (2005) Malarial parasites decrease reproductive success: an experimental study in a passerine bird. *Oecologia* 142:541–545.

Marzal A, Bensch S, Reviriego M, Balbontín J, de Lope F (2008) Effects of malaria double infection in birds: one plus one is not two. *J Evol Bio* 21:979–987.

Marzal A, García-Longoria L, Cárdenas Callirgos JM, Sehgal RN (2014) Invasive avian malaria as an emerging parasitic disease in native birds of Peru. *Biol Invasions*. doi: 10.1007/s10530-014-0718-x

Marzal A, Ricklefs RE, Valkiūnas G, et al. (2011) Diversity, loss, and gain of malaria parasites in a globally invasive bird. *PLoS One* 6:8.

Mauchamp A (1997) Threats from alien plant species in the Galápagos Islands. *Conserv Biol* 11:260–263.

Mauchamp A, Aldaz I, Ortiz E, Valdebenito H (1998) Threatened species, a re-evaluation of the status of eight endemic plants of the Galápagos. *Biodivers Conserv* 7:97–107.

Merino S, Moreno J, Jose J, Arriero E (2000) Are avian blood parasites pathogenic in the wild? A medication experiment in blue tits (*Parus caeruleus*). *R Soc* 9:2507–2510.

Merino S, Moreno J, Vásquez RA, et al. (2008) Haematozoa in forest birds from southern Chile: Latitudinal gradients in prevalence and parasite lineage richness. *Austral Ecol* 33:329–340.

Miller DFB, Holtzman SL, Kalkbrenner A, Kaufman TC (2001a) Homeotic Complex (Hox) gene regulation and homeosis in the mesoderm of the *Drosophila melanogaster* embryo: the roles of signal transduction and cell autonomous regulation. *Mech Dev* 102:17–32.

Miller GD, Hofkin BV, Snell H (2001b) Avian malaria and Marek's disease: potential threats to Galápagos penguins *Spheniscus mendiculus*. *Mar Ornithol* 29:43–46.

Nims BD, Vargas FH, Merkel J, Parker PG (2008) Low genetic diversity and lack of population structure in the endangered Galápagos penguin (*Spheniscus mendiculus*). *Conserv Genet* 9:1413–1420.

Obeyesekere G (1994) The apotheosis of Captain Cook: European Mythmaking in the Pacific. *Pac Aff* 67:159-162.

Olson S, James H (1982) Fossil birds from the Hawaiian Islands: evidence for wholesale extinction by man before Western contact. *Science* 217:633-635.

Padilla LR, Santiago-Alarcón D, Merkel J, Miller RE, Parker PG (2004) Survey for Haemoproteus spp., *Trichomonas gallinae*, *Chlamydophila psittaci*, and *Salmonella* spp. in Galápagos Islands columbiformes. *J Zoo Wildl Med* 35:60–64.

Palinauskas V, Kosarev V, Shapoval A, et al. (2007) Comparison of mitochondrial cytochrome b lineages and morphospecies of two avian malaria parasites of the subgenera *Haemamoeba* and *Giovannolaia* (Haemosporida: Plasmodiidae). *Zootaxa* 1626:39–50.

Palinauskas V, Valkiūnas G, Bolshakov CV, Bensch S (2008) *Plasmodium relictum* (lineage P-SGS1): effects on experimentally infected passerine birds. *Exp Parasitol* 120:372–380.

Paterson AM, Gray R (1997) *Host-parasite cospeciation, host switching, and missing the boat.* In Clayton & Moore (eds.): Host-Parasite Evolution: General Principles & Avian Models, Oxford University Press, Oxford, UK.

Phillips BL, Kelehear C, Pizzatto L, et al. (2010) Parasites and pathogens lag behind their host during periods of host range advance. *Ecology* 91:872–881.

Prenter J, MacNeil C, Dick JTA, Dunn AM (2004) Roles of parasites in animal invasions. *Trends Ecol Evol* 19:385–390.

Ricklefs RE (2010) Host-pathogen coevolution, secondary sympatry and species diversification. *Philos Trans R Soc Lond B Biol Sci* 365:1139–1147.

Ricklefs RE, Bermingham E (2007) The causes of evolutionary radiations in archipelagoes: Passerine birds in the Lesser Antilles. *Am Nat* 169:285–297.

Samuel M, Hobbelen P, Decastro F, et al. (2010) The dynamics, transmission, and population impacts of avian malaria in native Hawaiian birds: A modeling approach. *Ecol App* 21:2960–2973.

Santiago-Alarcón D, Outlaw DC, Ricklefs RE, Parker PG (2010) Phylogenetic relationships of haemosporidian parasites in New World Columbiformes, with emphasis on the endemic Galápagos dove. *Int J Parasitol* 40:463–470.

Santiago-Alarcón D, Whiteman NK, Parker PG, et al. (2008) Patterns of parasite abundance and distribution in island populations of Galápagos endemic birds. *J Parasitol* 94:584–590.

Schmid-Hempel P (2011) Evolutionary parasitology: The integrated study of infections, immunology, ecology and genetics. *Trends Parasitol* 27:233–234.

Schoener ER, Banda M, Howe L, Castro IC, Alley MR (2014) Avian malaria in New Zealand. *N Z Vet J* 62:189–198.

Schofield EK (1989) Effects of introduced plants and animals on island vegetation: Examples from the Galápagos Archipelago. *Conservation Biology* 3:227-238.

Sodhi N (2010) Birds. In Simberloff & Rejmanek (eds.) *Encyclopedia of Invasive Introduced Species. University* of California Press, California, USA.

Thomson GM (1922) *The Naturalisation of Animals and Plants in New Zealand.* Cambridge.

Tomás G, Merino S, Moreno J, Morales J (2007) Consequences of nest reuse for parasite burden and female health and condition in blue tits, *Cyanistes caeruleus*. *Anim Behav* 73:805–814.

Tompkins DM, Dunn AM, Smith MJ, Telfer S (2011) Wildlife diseases: From individuals to ecosystems. *J Anim Ecol* 80:19–38.

Tompkins DM, Gleeson DM (2006) Relationship between avian malaria distribution and an exotic invasive mosquito in New Zealand. *J R Soc New Zeal* 36:51–62.

Tompkins DM, Poulin R (2006) Parasites and Biological Invasions, Biological. *Biol Invasions New Zeal* 186:67–84.

Tourchin ME, Lafferty KD, Kuris AM (2002) Parasites and marine invasions. *Parasitology* 124:137–151.

Travis EK, Vargas FH, Merkel J, Gottdenker N, Miller RE, Parker PG (2006) Hematology, serum chemistry, and serology of Galápagos penguins (*Spheniscus mendiculus*) in the Galápagos Islands, Ecuador. *J. Wildl. Dis.* 42:625-632.

Valkiūnas G (2005) *Avian malaria parasites and other Haemosporidia*, Boca Raton.

Valkiūnas G, Tadas Z, Shapoval AP, Lezhova TA (2006) Effect of *Haemoproteus belopolskyi* (Haemosporida: Haemoproteidae) on body mass of the blackcap *Sylvia atricapilla*. *Am Soc Parasitol* 92:1123–1125.

Van Riper C, Van Riper SG, Goff ML, Laird M (1986) The epizootiology and ecological significance of malaria in Hawaiian land birds. *Ecol Monogr* 56:327–344.

Webster RG, Govorkova EA (2006) H5N1 Influenza — Continuing Evolution and Spread. *N Engl J Med* 355:2174–2177.

White EM, Wilson JC, Clarke AR (2006) Biotic indirect effects: A neglected concept in invasion biology. *Divers Distrib* 12:443–455.

Whiteman NK, Goodman SJ, Sinclair BJ, Walsh T, Cunningham AA, Kramer LD, Parker, PG (2005) Establishment of the avian disease vector *Culex quinquefasciatus* Say, 1823 (Diptera: *Culicidae*) on the Galápagos Islands, Ecuador. *Ibis* 147:843–847.

Wilmshurst JM, Anderson AJ, Higham TFG, Worthy TH (2008) Dating the late prehistoric dispersal of Polynesians to New Zealand using the commensal Pacific rat. *Proc Natl Acad Sci USA* 105:7676–7680.

In: Biological Invasions
Editor: Rebecca Waterman
ISBN: 978-1-63482-019-6

Chapter 3

AQUATIC ALIEN ANIMALS IN CHINA: THEIR INTRODUCTION, INVASION AND MANAGEMENT

Du Luo*[1], *Meng Xu*[2], *Xidong Mu*[2], *Dangen Gu*[1], *Hui Wei*[2], *Yexin Yang*[2] *and Yinchang Hu*

[1]Key Laboratory of the Conservation and Ecological Restoration of Fishery Resource in Pearl River, Pearl River Fisheries Research Institute, Chinese Academy of Fishery Sciences, Guangzhou, China
[2]Key Laboratory of Tropical & Subtropical Fishery Resource Application & Cultivation, Ministry of Agriculture, Pearl River Fisheries Research Institute, Chinese Academy of Fishery Sciences, Guangzhou, China

ABSTRACT

Globally, biological invasion is a threat to economy, biodiversity, and human health. Most of the alien species were mediated by human activity, which is true for aquatic invasion species as well. China has become the world's second largest economic entity and the world's first aquaculture and fisheries trade powers. A total of 544 alien invasive species, including 51 of 100 of the world's worst invasive alien species,

* Corresponding author: Pearl River Fisheries Research Institute, Chinese Academy of Fishery Sciences, 1st Xingyu Road, Liwan District, Guangzhou City, Guangdong Province, China. E-mail address: huyc22@163.com (Yin-Chang Hu), Tel.: +86 20 81616088; fax: +86 20 81616162 (work).

have successfully invaded China and have caused great damage to ecological environment, economic development, and public health; ultimately caused by the increase of trade, tourism and logistics. To date, however, few systematic reviews existed to summarize the process, impact and management measures concerning invasive species, especially for aquatic non-native invasive species in China. Here in this chapter, we first reviewed the advance of aquatic animal invasion, including the definition of aquatic animal invasion and their unique presence in the world biological invasion. Secondly, we listed 31 main invasive and potential nuisance aquatic alien animal species, 221introduced species including 100 aquaculture animals and 121 ornamental fish, summarized the ecological impact, economic loss and invasion progress, and gave some examples with respect to ecologic destruction. Thirdly, taking apple snail (*Pomacea canaliculata*) and fish as examples, we comprehensively analyzed the management strategy, and the current prevention and control measures for invasive alien aquatic animals. Finally, we gave some suggestions about the future management and research topics of aquatic animal invasion.

Keywords: Aquatic invasion, biodiversity conservation, introduced aquaculture species, nonnative ornamental fish, management strategy

INTRODUCTION

Globally, biological invasion is a threat to the economy, biodiversity, and human health (Levine 2000). Species invasion is one of the leading mechanisms of global environmental change, particularly in freshwater ecosystems (García-Berthou et al. 2005) and has become one of the five most important threats to global freshwater biodiversity (Dudgeon et al. 2006). By various pathways, the exotic aquatic invertebrates, fishes, and plant species have been introduced, become established, and finally invasively caused significant damage to the local ecosystems, to the economies, to human health, and to native biodiversity. The increasing introduction and dispersal of invasive species in freshwater and marine environments is a worldwide problem that is being given more and more attention (Strayer 2010).

Most of the alien species were introduced through human activity. China has become the world's second largest economic entity and the world’s first aquaculture and fisheries trade powers. The introduction and invasion of aquatic animals into China was in line with the increase of the gross domestic production of the previous 35 years in Eastern Asia. With the development of society and economics, and the increase of trade, tourism and logistics, the damage caused by the invasion of aquatic organisms greatly increased (Xu et al. 2006). Currently, 51 of 100 of the world's worst invasive alien species have successfully invaded into China. Over 200 exotic species, not including subspecies and hybrids, are maintained in various kinds of waters in mainland China and the number is still growing. How to manage the increasingly aquatic biological invasions in China under globalization? It is a complex issue requiring tremendous long-term efforts, scientific technology, and wide cooperation from both the government and the masses (Hulme 2009; Perrings et al. 2005).

Biological invasion has not only threatened the biological diversity but has also affected the ecological environment and economic development. Successful invaders can in many cases inflict enormous ecological damage through predation, grazing, competition, habitat alteration, hybridization, resistance to drugs, and disease transportation (Mack et al. 2000). In order to clarify the aquatic invasion status, and to promote aquatic invasion management, and aquatic biodiversity protection, we here reviewed the advance of alien invasive aquatic animals and their management in China by clarifying definition, listing the alien invasive species (giving scientific names according to FishBase: http://www.fishbase.org and Integrated Taxonomic Information System, ITIS: http://www.itis.gov), analyzing their origin, current distribution and impact, discussing the status of related laws and regulations, and giving some successful examples related to successful prevention and control of invasive aquatic animal species.

Definition of Aquatic Animal Invasion

The publication of Charles Elton's book, "The Ecology of Invasions by Animals and Plants" in 1958, launched the systematic study of biological invasions. Nowadays, invasion ecology has grown to become an important multi-disciplinary subfield of ecology with growing links to many other disciplines (Richardson & Pysek 2008). The study of biological invasions has triggered the production of a variety of concepts (Falk-Petersen et al. 2006).

During a biological invasion, the alien species would firstly acquire competitive advantages following the disappearance of natural obstacles to their proliferation, which allow themselves to spread rapidly. Then, they conquer novel areas and become a dominant population within the recipient ecosystems. Finally, the exotic species changed into invasive organisms (Valery et al. 2008). Thus, based on the foregoing considerations, invasion ecology is the study of the establishment, spread, and ecological impact of species translocated from one region or continent to another by humans. Biological invasions have been used as a tool for basic ecological research to study the ecology and evolution of populations and of novel biotic interactions, and the conservation of biodiversity (Sakai et al. 2001).

Aquatic invasion means biological invasion of aquatic species including introduced aquatic exotic flora and fauna. Aquatic invasive species are non-native species whose introduction does or is likely to cause economic, environmental or health harm, specifically threaten the biodiversity of native species and the ecological stability of infested waters, commercial and agricultural production activities that dependent on such waters, or public health by carrying pathogens and resulting in disease.

Accordingly, aquatic animal invasion is defined as aquatic animal species' invasion. In invasive aquatic animal species, there is a distinction between the introduced species meaning any aquatic species growing outside its natural habitat with many cultivated or domesticated organisms, and invasive species referring to the introduced aquatic species that establish free-living populations in the natural water and caused some damage to ecology or the economy. Invasive aquatic animals require a watery habitat, but do not necessarily have to live entirely in water. Only a small minority of those introduced species, such as apple snail (*Pomacea canaliculata*), zebra mussel (*Dreissena polymorpha*), and water hyacinth (*Eichhornia crassipes*), become invaders with certain problems of economic and ecological costs. The other introduced species are generally accepted as beneficial additions, such as most major food animals. The terms exotic, alien, introduced, non-indigenous, or non-native species all refer to plants, animals, or microscopic organisms living outside their native range. And, aquatic invasive species and aquatic nuisance species are synonymous (Congress 1990).

INTRODUCED ALIEN AQUATIC ANIMALS IN CHINA

The invasive alien species are introduced by lots of pathways, by which invasive species are imported or introduced into new environments. Natural pathways like wind and water currents account for very few of the new species introductions. Humans caused the vast majority of alien species introductions (Mack et al. 2000). Increased globalization of long-distance travel and commerce has increased the frequency of introductions of non-native animals to aquatic ecosystems in China (Yan et al. 2001). The transport of aquatic invasive species has been identified as seven major pathways, which are shipping, recreational and commercial boating, live bait, aquarium and water garden trade, live food fish, unauthorized or authorized introductions, and canals and water diversions.

Generally, there are three invasion patterns according to the difference of human-aided pathways. Firstly, a transportation-related shipment or waterway may lead to the dispersal of aquatic species. At least five classes of lower aquatic animals were detected in the ballast water of exotic ships entering major ports of Fujian province (Yang et al. 2009). Secondly, ill-advised or consuetudinary intentional release is still an important source of invasion. These releases include government-organized activities, such as the release of mosquito fish (*Gambusia affinis*), and the activity of 'free captive animals' based on the folk custom or the inappropriate public education. Thirdly, accidental or ill-planned release is the main source of invasive species and potential nuisance species. There are two kinds of this release, besides the escaping of aquatic animals from science laboratories and zoos. One is the introduction of aquatic species for aquaculture to increase food production. About 100 aquatic non-native animal species (Table 1) have been introduced, and multiply introduced into China in the past four decades; most of them were carried out in the 1980s and 1990s. The other is aquarium trade in which abundant of species have been imported, transported, sold, exchanged, farmed and sometimes released privately. Most of the ornamental animals are fishes mainly from the tropical and subtropical areas of Southeast Asia, South America, and East Africa. Here in table 2 we listed the popular ornamental fish species that lives with some scale in Mainland China. They are mainly classified into three families of Cyprinidae, Characidae, and Cichlidae. In fact, apart from the above listed fish species, there are a lot more subspecies and hybrid fish. For example, the blood parrot (*Paraneetroplus synspilus*× *Amphilophus citrinellus*), red tilapia (*Oreochromis niloticus* × *Oreochromis mossambicus*), and striped bass (*Morone saxatilis*♀ × *Morone chrysops*♂) are all hybrids with a large farming amount. Many specific varieties are continually being developed by breeding methods of cultivation and hybridization.

Table 1. List of the introduced non-native animals for aquaculture in China

No.	Scientific name	Chinese name	Category	Origin	Time
Mollusca					
1	*Pomacea canaliculata*	福寿螺	Ampullariidae	Argentina	1981
2	*Pomacea insularum*	福寿螺	Ampullariidae	Argentina	1981
3	*Haliotis discus*	日本盘鲍	Haliotidae	Japan	1986
4	*Haliotis fulgens*	绿鲍	Haliotidae	USA	1984
5	*Haliotis gigantea*	日本大鲍	Haliotidae	Japan	1997
6	*Haliotis laevigata*	绿唇鲍	Haliotidae	Australia	1998
7	*Haliotis rufescens*	红鲍	Haliotidae	USA	1984
8	*Panopea generosa*	太平洋潜泥蛤	Hiatellidae	North America	1998
9	*Crassostrea gigas*	太平洋牡蛎	Ostreidae	Japan	1979
10	*Argopecten irradians*	海湾扇贝	Pectinidae	USA	1981
11	*Patinopecten yessoensis*	虾夷扇贝	Pectinidae	Japan	1980
12	*Pecten maximus*	欧洲大扇贝	Pectinidae	Europe	1997
13	*Mercenaria mercenaria*	硬壳蛤	Veneridae	USA	1997
Arthropod					
14	*Procambarus darkii*	克氏原鳌虾	Cambaridae	Australia	1990
15	*Macrobrachium rosenbergii*	罗氏沼虾	Palaemonidae	Southeast Asia	1976
16	*Cherax quadricarinatus*	红螯鳌虾	Parastacidae	Australia	1990
17	*Cherax tenuimanus*	麦龙鳌虾	Parastacidae	Australia	1985
18	*Cherax destructor*	亚比虾	Parastacidae	Australia	1995
19	*Metapenaeus ensis*	刀额新对虾	Penaeidae	Australia	1992
20	*Penaeus monodon*	斑节对虾	Penaeidae	Thailand	1986
21	*Penaeus stylirostris*	南美蓝对虾	Penaeidae	South America	2000
22	*Penaeus vannamei*	南美白对虾	Penaeidae	South America	1988
Echinoderm					
23	*strongylocentrotus intermedius*	虾夷马粪海胆	Strongylocentrotidae	Japan	1989

No.	Scientific name	Chinese name	Category	Origin	Time
Urochordata					
24	*Halocynthia roretzi*	真海鞘	Pyuridae	Japan, Korea	2006
Vertebrata: Fish					
25	*Acipenser baerii*	西伯利亚鲟	Acipenseridae	Siberia	Late 90s
26	*Acipenser nudiventris*	裸腹鲟	Acipenseridae	Europe	Late 90s
27	*Acipenser ruthenus*	小体鲟	Acipenseridae	Russia	1995
28	*Acipenser sturio*	大西洋鲟	Acipenseridae	Eastern Atlantic	Late 90s
29	*Acipenser gueldenstae dtii*	俄罗斯鲟	Acipenseridae	Russia	1997
30	*Angullla angullla*	欧洲鳗	Anguillidae	Europe	1993
31	*Anguilla rostrata*	美洲鳗	Anguillidae	North America	1998
32	*Ictiobus cyprinellus*	大口胭脂鱼	Catostomidae	North America	1997
33	*Lepomis cyanellus*	蓝太阳鱼	Centrarchidae	North America	Late 90s
34	*Lepomis gibbosus*	驼背太阳鱼	Centrarchidae	North America	1998
35	*Lepomis macrochirus*	蓝鳃太阳鱼	Centrarchidae	North America	1987
36	*Pomoxis nigromaculatus*	黑莓鲈	Centrarchidae	USA	1993
37	*Micropterus salmoides*	大口黑鲈	Centrarchidae	USA	1983
38	*Channa striata*	纹鳢	Channidae	Southeast Asia	1991
39	*Oreochromis aureus*	奥利亚罗非鱼	Cichlidae	Africa	1981
40	*Oreochromis mossambicus*	莫桑比克罗非鱼	Cichlidae	Africa	1957
41	*Oreochromis niloticus*	尼罗罗非鱼	Cichlidae	Africa	1978
42	*Sarotherodon galilaeus*	加利亚罗非鱼	Cichlidae	Africa	Late 80s
43	*Tilapia zillii*	齐氏罗非鱼	Cichlidae	Africa	Late 80s
44	*Clarias batrachus*	蟾胡子鲶	Clariidae	Thailand	1978
45	*Clarias macrocephalus*	斑点胡子鲶	Clariidae	Thailand	1982

Table 1. (Continued)

No.	Scientific name	Chinese name	Category	Origin	Time
46	*Clarias gariepinus*	革胡子鲶	Clariidae	Egypt	1981
47	*Pangasianodon gigas*	巨鲶	Clariidae	Southeast Asia	1986
48	*Carassius cuvieri*	日本白鲫	Cyprinidae	Japan	1976
49	*Catla Catla*	卡特拉鲃	Cyprinidae	South Asi a	1982
50	*Cirrhinus mrigala*	麦穗加拉鲮鱼	Cyprinidae	Southeast Asia	1982
51	*Cyprinus carpio carpio*	德国镜鲤	Cyprinidae	Germany	1984
52	*Labeo rohita*	露斯塔野鲮	Cyprinidae	Southeast Asia	1978
53	*Leptobarbus hoevenii*	何氏细须鲃	Cyprinidae	Thailand	1986
54	*Puntius gonionotus*	银鲃	Cyprinidae	Southeast Asia	1986
55	*Tinca tinca*	丁鱥	Cyprinidae	Europe	1998
56	*Oxyeleotris marmorata*	云斑尖塘鳢	Eleotridae	Southeast Asia	1987
57	*Embiotoca jacksoni*	海鲫	Embiotocidae	USA	1997
58	*Rhinogobius brunneus*	褐吻鰕虎鱼	Gobiidae	Thailand	1986
59	*Ameiurus nebulosus*	云斑鮰	Ictaluridae	USA	1978
60	*Ictalurus punctatus*	斑点叉尾鮰	Ictaluridae	USA	1984
61	*Lates niloticus*	尖吻鲈	Latidae	Egypt	1985
62	*Morone saxatilis*	条纹石鮨	Moronidae	Western Atlantic	1993
63	*Morone chrysops*	白鲈	Moronidae	USA	1998
64	*Morone saxatilis*	条纹鲈	Moronidae	USA	1991
65	*Pangasianodon hypophthalmus*	苏氏鲶	Pangasiidae	Southeast Asia	1978
66	*Paralichthys dentatus*	大西洋牙鲆	Paralichthyidae	USA	2002
67	*Paralichthys lethostigma*	漠斑牙鲆	Paralichthyidae	USA	2001
68	*Macquaria ambigua*	澳洲金鲈	Percichthyidae	Australia	1991
69	*Sander vitreus*	大眼狮鲈	Percidae	North America	1993
70	*Verasper moseri*	条斑星鲽	Pleuronectidae	Japan	2004
71	*Polyodon spathula*	匙吻鲟	Polyodontidae	USA	1995

No.	Scientific name	Chinese name	Category	Origin	Time
72	*Prochilodus lineatus*	巴西鲷	Prochilodontidae	Brazil	1996
73	*Prochilodus scrofa*	小口脂鲤	Prochilodontidae	South America	1998
74	*Coregonus autumnalis*	凹目白鲑	Salmonidae	Russia	1997
75	*Coregonus heglingus*	奇白鲑	Salmonidae	Russia	1987
76	*Coregonus maraenoides*	楚德白鲑	Salmonidae	Russia	1985
77	*Coregonus peled*	高白鲑	Salmonidae	Russia	1985
78	*Oncorhynchus kisutch*	银鲑	Salmonidae	USA	1987
79	*Oncorhynchus mykiss*	虹鳟	Salmonidae	USA	1983
80	*Salmo salar*	大西洋鲑	Salmonidae	USA	2000
81	*Sciaenops Ocellatus*	眼斑拟石首鱼	Sciaenidae	USA	1991
82	*Scophthalmus maximus*	大菱鲆	Scophthalmidae	UK	1992
83	*Colossoma brachypomum*	短盖巨脂鲤	Serrasalmidae	South America	1985
84	*Silurus glanis*	欧洲六须鲶	Siluridae	Europe	Late 90s
85	*Solea solea*	欧鳎	Soleidae	Europe	2003
86	*Sparus aurata*	金头鲷	Sparidae	France	2001
87	*Bidyanus bidyanus*	澳洲银鲈	Terapontidae	Australia	1991
Vertebrata: Amphibians and Reptiles					
88	*Chelydra serpentina*	拟鳄龟	Chelydridae	USA	1999
89	*Macrochelys temminckii*	大鳄龟	Chelydridae	USA	1999
90	*Crocodylus porosus*	湾鳄	Crocodylidae	Southeast Asia	1987
91	*Chrysemys picta*	锦龟	Emydidae	USA	1993
92	*Trachemys scripta elegans*	红耳彩龟	Emydidae	USA	80s
93	*Trachemys scripta scripta*	黄耳彩龟	Emydidae	USA	90s
94	*Geoemyda japonica*	日本地龟	Geoemydidae	Japan	1993
95	*Lithobates heckscheri*	河蛙	Ranidae	USA	1987
96	*Rana catesbeiana*	牛蛙	Ranidae	Cuba	1959
97	*Rana grylio*	美国青蛙	Ranidae	USA	1987

Table 1. (Continued)

No.	Scientific name	Chinese name	Category	Origin	Time
98	*Apalone ferox*	珍珠鳖	Trionychidae	USA	1993
99	*Apalone mutica*	美国鳖	Trionychidae	USA	1993
100	*Dogania subplana*	马来鳖	Trionychidae	Malaysia	90s

Table 2. List of the non-native ornamental fish species in China

No.	Scientific name	Chinese name	Category	Origin	Time
Poeciliidae:					
1	*Poecilia latipinna*	宽帆鳉	Poeciliidae	Mexico, USA	90s
2	*Poecilia reticulata*	花鳉	Poeciliidae	Venezuela, Guyanas	80s
3	*Xiphophorus helleri*	剑尾鱼	Poeciliidae	Mexico, Guatemala	80s
4	*Xiphophorus maculatus*	斑剑尾鱼	Poeciliidae	Mexico, Guatemala	80s
Anablepidae:					
5	*Anableps anableps*	四眼鱼	Anablepidae	South America	90s
Cyprinidae:					
6	*Akrokolioplax bicornis*	角鱼	Cyprinidae	Indonesia	90s
7	*Barbus dorsalis*	四间鲃	Cyprinidae	Malaysia, Indonesia	90s
8	*Barbonymus schwanenfeldii*	多鳞四须鲃	Cyprinidae	Thailand, Indonesia	90s
9	*Barbodes lateristriga*	侧纹四须鲃	Cyprinidae	Southeast Asia	80s
10	*Balantiocheilus melanopterus*	黑鳍袋唇鱼	Cyprinidae	Thailand, Indonesia	90s
11	*Cyprinus carpio*	日本锦鲤	Cyprinidae	Japan	80s
12	*Danio malabaricus*	大斑马鱼	Cyprinidae	India	90s
13	*Danio rerio*	斑马鱼	Cyprinidae	India, Bangladesh	80s
14	*Epalzeorhynchos bicolor*	双色野鲮	Cyprinidae	Thailand	90s
15	*Epalzeorhynchos frenatus*	彩虹鲨	Cyprinidae	Thailand	90s

No.	Scientific name	Chinese name	Category	Origin	Time
16	*Epalzeorhynchos kalopterus*	高体波鱼	Cyprinidae	Southeast Asia	90s
17	*Labeo chrysophekadion*	黑野鲮	Cyprinidae	Southeast Asia	90s
18	*Oreichthys cosuatis*	山鲋	Cyprinidae	Indonesia	90s
19	*Puntius conchonius*	玫瑰无须鲃	Cyprinidae	India	90s
20	*Puntius sachsii*	黄金鲃	Cyprinidae	Singapore	90s
21	*Puntius titteya*	樱桃无须鲃	Cryprinidae	Sri Lanka	90s
22	*Sahyadria denisonii*	丹尼氏无须鲃	Cyprinidae	India	21th
23	*Trigonopoma pauciperforatum*	红线波鱼	Cyprinidae	Malaysia, Indonesia	90s
Ambassidae:					
24	*Parambassis ranga*	兰副双边鱼	Ambassidae	Southeast Asia	90s
Characidae:					
25	*Gymnocorymbus te rnetzi*	裸顶脂鲤	Characidae	South America	80s
26	*Hasemania nana*	银顶光尾裙鱼	Characidae	Brazil	90s
27	*Hemigrammus ocellifer*	眼斑半线脂鲤	Characidae	Amazon River	90s
28	*Hemigrammus rhodostomus*	红吻半线脂鲤	Characidae	Amazon River	90s
29	*Hemigrammus rodwayi*	罗氏半线脂鲤	Characidae	South America	90s
30	*Hyphessobrycon eques*	丽鲃脂鲤	Characidae	Amazon	90s
31	*Hyphessobrycon erythrostigma*	红点鲃脂鲤	Characidae	Columbia	90s
32	*Hyphessobrycon flammeus*	火焰鲋脂鲤	Characidae	South America	90s
33	*Hyphessobrycon pulchripinnis*	丽鳍鲋脂鲤	Characidae	South America	90s
34	*Hyphessobrycon rosaceus*	玫瑰鲋脂鲤	Characidae	South America	90s
35	*Moenkhausia sanctaefilomenae*	红目脂鲤	Characidae	South America	90s

Table 2. (Continued)

No.	Scientific name	Chinese name	Category	Origin	Time
36	*Paracheirodon axelrodi*	唇齿脂鲤	Characidae	Amazon River	90s
37	*Paracheirodon innesi*	拟唇齿脂鲤	Characidae	Amazon River	80s
38	*Prionobrama filige ra*	血鳍玻璃鱼	Characidae	South America	90s
Serrasalmidae:					
39	*Metynnis hypsauchen*	舒氏银脂鲤	Serrasalmidae	Guiana, Paraguay	90s
40	*Piaractus brachyp omus*	短盖巨脂鲤	Serrasalmidae	Amazon River	80s
41	*Pygocentrus nattereri*	纳氏距脂鲤	Serrasalmidae	Amazon River	21th
Lebiasinidae:					
42	*Nannostomus harrisoni*	哈氏铅笔鱼 金铅笔鱼	Lebiasinidae	Guyana	90s
43	*Nannostomus marginatus*	短铅笔鱼	Lebiasinidae	South America	90s
44	*Nannostomus trifasciatus*	三带铅笔鱼	Lebiasinidae	South America	90s
Alestidae:					
45	*Phenacogrammus interruptus*	断点脂鲤	Alestidae	Africa	21th
Anostomidae:					
46	*Anostomus anostomus*	红尾上口脂鲤	Anostomidae	South America	21th
47	*Leporinus affinis*	安芬兔脂鲤	Anostomidae	South America	90s
Gasteropelecidae:					
48	*Gasteropelecus sternicla*	胸斧鱼	Gasteropelecidae	Amazon River	80s
Prochilodontidae					
49	*Semaprochilodus insignis*	旗尾真唇脂鲤	Prochilodontidae	South America	90s
Cichlidae:					
50	*Amatitlania nigrofasciata*	九间丽体鱼	Cichlidae	Central America	21th
51	*Amphilophus citrinellus*	橘色双冠丽鱼	Cichlidae	Central America	80s

No.	Scientific name	Chinese name	Category	Origin	Time
52	*Andinoacara pulcher*	蓝宝丽鱼	Cichlidae	Trinidad, Venezuela	90s
53	*Andinoacara rivulatus*	绿宝丽鱼	Cichlidae	Ecuador, Peru	90s
54	*Apistogramma agassizii*	阿氏隐带丽鱼	Cichlidae	South America	21th
55	*Apistogramma brevis*	短身隐带丽鱼	Cichlidae		21th
56	*Apistogramma guttata*	斑点隐带丽鱼	Cichlidae	Venezuela	21th
57	*Apistogramma pleurotaenia*	侧带隐带丽鱼	Cichlidae	Argentina	21th
58	*Apistogramma panduro*	壮身隐带丽鱼	Cichlidae	Peru, Brazil	21th
59	*Apistogrammoides pucallpaensis*	普卡尔似隐带丽鱼	Cichlidae	Peru	90s
60	*Astronotus ocellatus*	图丽鱼	Cichlidae	South America	90s
61	*Cichlasoma trimaculatum*	三斑丽体鱼	Cichlidae	Central America	21th
62	*Copadichromis azureus*	阿祖桨鳍丽鱼	Cichlidae	Lake Malawi	90s
63	*Copadichromis chrysonotus*	金桨鳍丽鱼	Cichlidae	Lake Malawi	21th
64	*Hemichromis bimaculatus*	双斑伴丽鱼	Cichlidae	West Africa	90s
65	*Hemichromis guttatus*	点纹伴丽鱼	Cichlidae	Congo River in Africa	90s
66	*Hemichromis lifalili*	玫瑰伴丽鱼	Cichlidae	Congo River basin	90s
67	*Herichthys cyanoguttatus*	蓝斑丽体鱼	Cichlidae	Texas, Mexico	90s
68	*Heros severus*	庄严丽体鱼	Cichlidae	Amazon River basin	90s
69	*Labidochromis caeruleus*	淡黑镊丽鱼	Cichlidae	Lake Malawi	90s
70	*Melanochromis auratus*	纵带黑丽鱼	Cichlidae	Lake Malawi	90s

Table 2. (Continued)

No.	Scientific name	Chinese name	Category	Origin	Time
71	*Mesonauta festivus*	花边中丽鱼	Cichlidae	South America	90s
72	*Mikrogeophagus altispinosus*	高棘小噬土丽鲷	Cichlidae	Bolivia	90s
73	*Mikrogeophagus ramirezi*	拉氏小噬土丽鲷	Cichlidae	Venezuela	90s
74	*Nanochromis nudiceps*	裸头彩短鲷	Cichlidae	Congo River basin	90s
75	*Parananochromis caudifasciatus*	尾纹副南丽鱼	Cichlidae	Africa	90s
76	*Paraneetroplus synspilus*	红头丽体鱼	Cichlidae	Central America	80s
77	*Pelvicachromis taeniatus*	带纹矛耙丽鱼	Cichlidae	Western Africa	21th
78	*Pterophyllum altum*	横纹神仙鱼	Cichlidae	South America	21th
79	*Pterophyllum leopoldi*	利氏神仙鱼	Cichlidae		
80	*Pterophyllum scalare*	大神仙鱼	Cichlidae	South America	21th
81	*Pterophyllum scalare*	大神仙鱼	Cichlidae	Amazon River	80s
82	*Symphysodon aequifasciatus*	盘丽鱼	Cichlidae	South America	90s
83	*Taeniacara candidi*	坎氏纹首丽鱼	Cichlidae	South America	21th
84	*Thorichthys meeki*	火口鱼	Cichlidae	Central America	90s
85	*Tilapia buttikoferi*	布氏非鲫	Cichlidae	Western Africa	90s
Datnioididae:					
86	*Datnioides microlepis*	小鳞拟松鲷	Datnioididae	Southeast Asia	90s
Pomacentridae:					
87	*Premnas biaculeatus*	棘颊雀鲷	Pomacentridae	Indo-West Pacific	90s
Monodactylidae:					

No.	Scientific name	Chinese name	Category	Origin	Time
88	*Monodactylus sebae*	油脂大眼鲳	Monodactylidae	Eastern Atlantic	90s
Osphronemidae:					
89	*Betta splendens*	五彩搏鱼	Osphronemidae	Mekong basin	80s
90	*Osphronemus goramy*	丝足鲈	Osphronemidae	Southeast Asia	90s
91	*Trichogaster lalius*	拉利毛足鲈	Osphronemidae	India	90s
92	*Trichopodus leerii*	珍珠毛足鲈	Osphronemidae	Southeast Asia	80s
93	*Trichopodus trichopterus*	丝鳍毛足鲈	Osphronemidae	Southeast Asia	80s
Helostomatidae:					
94	*Helostoma temminkii*	吻鲈	Helostomatidae	Thailand to Indonesia	80s
95	Tetraodontidae:				
96	*Carinotetraodon travancoricus*	狡鲀	Tetraodontidae	India	21th
Notopteridae:					
97	*Chitala chitala*	铠甲弓背鱼	Notopteridae	India	90s
98	*Chitala blanci*	白氏铠甲弓背鱼	Notopteridae	Mekong basin	90s
99	*Chitala lopis*	大眼铠甲弓背鱼	Notopteridae	Mekong basin	90s
100	*Notopterus notopterus*	弓背鱼	Notopteridae	Thailand, Malaysia	90s
101	*Papyrocranus afer*	非洲驼背鱼	Notopteridae	Africa	90s
102	*Xenomystus nigri*	光背鱼	Notopteridae	Africa	90s
Gyrinocheilidae:					
104	*Gyrinocheilus aymonieri*	湄公双孔鱼	Gyrinocheilidae	Southeast Asia	90s
Loricariidae:					
105	*Hypostomus plecostomus*	下口鲶	Loricariidae	South America	90s
Pimelodidae:					
106	*Pseudoplatystoma fasciatum*	条纹鸭嘴鲶	Pimelodidae	South America	90s
107	*Phractocephalus hemiliopterus*	红尾护头鲿	Pimelodidae	South America	90s
Pangasiidae:					

Table 2. (Continued)

No.	Scientific name	Chinese name	Category	Origin	Time
108	*Pangasianodon hypophthalmus*	低眼无齿鲶	Pangasiidae	Southeast Asia	80s
Cobitidae:					
109	*Chromobotia macracanthus*	皇冠沙鳅	Cobitidae	Indonesia	21th
Siluridae:					
110	*Kryptopterus bicirrhis*	双须缺鳍鲶	Siluridae	Southeast Asia	80s
Osteoglossidae:					
111	*Osteoglossum bicirrhosum*	双须骨舌鱼	Osteoglossidae	South America	90s
112	*Osteoglossum ferreirai*	费氏骨舌鱼	Osteoglossidae	South America	90s
113	*Scleropages formosus*	美丽硬骨舌鱼	Osteoglossidae	Southeast Asia	90s
114	*Scleropages leichardti*	澳洲硬骨舌鱼	Osteoglossidae	Australia	21th
Arapaimidae:					
115	*Arapaima gigas*	巨巴西骨舌鱼	Arapaimidae	South America	21th
116	*Heterotis niloticus*	尼罗异耳骨舌鱼	Arapaimidae	Africa	21th
Lepisosteidae:					
117	*Lepisosteus osseus*	雀鳝	Lepisosteidae	North and Central America	90s
Potamotrygonidae:					
118	*Potamotrygon falkneri*	卡氏江魟	Potamotrygonidae	South America	21th
119	*Potamotrygon motoro*	南美江魟	Potamotrygonidae	South America	21th
120	*Potamotrygon schroederi*	施氏江魟	Potamotrygonidae	South America	21th
Mormyridae:					
121	*Gnathomemus petersi*	彼氏锥颌象鼻鱼	Mormyridae	Africa	90s

In the history of invasion study, the ecologists paid more attention to the adaptive (Elton 1958). The introduced food fish, because of their selection standard of good adaptability, strong ingestion and digestion, fast growth and large-scale popularization, have easily become and are becoming invasive aquatic species. For aquarium species, most of them are carnivorous and multiple spawning. To sum up, there are over 220 non-native aquatic animals in Mainland China in species levels apart from other ornamental animals, such as *Pomacea diffusa* and *Planorbarius corneus*, and dozens of subspecies and hybrids. Mosquito fish (*Gambusia affinis*), which is neither food fish nor ornamental fish, are a great threat to the survival of native invertebrates, fish and amphibians (Chen et al. 2008). The red-eared slider (*Trachemys scripta*), used both as food species and ornamental species, has a distribution of 22 provinces in natural water nearly all over China (Liu et al. 2011). Additionally, these species have a wide distribution of geographical origin, category, and environmental suitability. Obviously, it will be a heavy burden to the natural aquatic ecosystems in the future if we do not take measures to make this tendency better to protect biodiversity from biological invasion.

MAIN AQUATIC INVASIVE SPECIES AND IMPACTS

Three decades is a long enough period for the introduced invasive species to cause various environmental and ecological problems. In recent years, more and more alien invasive species, such as water hyacinth (*Eichhornia crassipes*), giant African snail (*Achating fulica*), apple snail (*Pomacea canaliculata*), loblolly pine mealybug (*Oracella acuta*), red imported fire ant (*Solenopsis invicta*), Amazon sailfin catfish (*Pterygoplichthys pardalis*), bull frog (*Rana catesbeiana*), etc., have seriously impaired the biological diversity and ecological environment of China, caused gross economic losses, and brought great social impact through some hot news of harm to people's health (Ding et al. 2008). In order to prevent invasive species, to protect biological diversity and ecological systems, to safeguard the environmental and social security, and finally to promote the sustainable development of China, the State Environmental Protection Administration of China and Chinese Academy of Sciences have formulated and issued three batches of the alien invasive species in China in 2003, 2010, and 2014 respectively (Table 3-5) (State Environmental Protection Administration of China 2003, 2010, 2014).

Table 3. First batch of alien invasive species list in China(2003)

No.	Scientific name	English name	Chinese name	Category	Origin	Invasion history	Current distribution
1	*Eupatofium adenophorum* Spreng	Crofton Weed	紫茎泽兰	Compositae	Central America	1935, Yunnan	Yunnan, Guangxi, Guizhou, Sichuan (Southwest), Taiwan
2	*Mikaina micrantha* H. B. K.	Mile-a-minute Weed	薇甘菊	Compositae	Central America	1884, Hong Kong	Hong Kong, Macao, and the Pearl River Delta of Guangdong Province
3	*Alternanthera philoxeroides* (Mart.) Griseb	Alligator Weed	空心莲子草	Amaranthaceae	South America	1892, Shanghai	Almost all over the southern areas up to the Yellow River of China, Tianjin
4	*Ambrosia artemisiifolia* L.	Ragweed, Bitterweed	豚草	Compositae	North America	1935, Hangzhou	About 15 provinces and municipalities under the Central Government in the Southeast, North, Middle, and East of China.
5	*Lolium temulentum* L.	Darnel Rye-grass, Poison Darnel	毒麦	Gramineae	Mediterranean area of the Europe	1954, first found in the wheat imported from Bulgaria	Presence has been reported by every province (region) except Tibet and Taiwan

No.	Scientific name	English name	Chinese name	Category	Origin	Invasion history	Current distribution
6	*Spartina alterniflora* Loisel.	Smooth Cord-grass	互花米草	Gramineae	Southeast coasts of the United States	It was introduced in 1979 and gained some economic benefits.	Shanghai, Zhejiang, Fujian, Guangdong, Hong Kong.
7	*Eupatorium odoratum* L.	Fragrant Eupatorium, Bitter Bush, Siam Weed	飞机草	Compositae	Central America	1934, Yunnan	Taiwan, Guangdong, Hong Kong, Macao, Hainan, Guangxi, Yunnan, and Guizhou.
8	*Eichhornia crassipes* (Mart.) Solms	Water Hyacinth	凤眼莲	Pontederiaceae	Northeast of Brazil	1901, Taiwan	Cultivated in 19 provinces (autonomous regions, and municipalities directly under the Central Government) in south Liaoning, North China, East China, Middle China, and South

Table 3. (Continued)

No.	Scientific name	English name	Chinese name	Category	Origin	Invasion history	Current distribution
							China, and is a wild grass in Yangtze River Area and the southern areas thereto.
9	*Sorghum halepense* (L.) Pers.	Johnson Grass	假高粱	Gramineae	Mediterranean area	Early 20th, Taiwan, Hong Kong, north Guangdong	Taiwan, Guangdong, Guangxi, Hainan, Hong Kong, Fujian, Hunan, Anhui, Jiangsu, Shanghai, Liaoning, Beijing, Hebei, Sichuan, Chongqing, and Yunnan.
10	*Opogona sacchari* (Bojer)	Banana Moth	蔗扁蛾	Lepidoptera, Hieroxestidae	Tropical and subtropical areas of Africa	1987, Guangzhou; 1990, Beijing	Spread in over 10 provinces and municipalities directly under the Central Government. The situation is more serious in southern regions, banana moth problems exist

No.	Scientific name	English name	Chinese name	Category	Origin	Invasion history	Current distribution
							wherever there are Dracaena fragrans Ker-Gawl in those places.
11	*Oracella acuta* (Lobdell)	Lobdelly Pine Mealybug	湿地松粉蚧	Homoptera, Pseudococcidae	The United States	1988, Taishan, Guangdong	Guangdong, Guangxi, and Fujian, etc.
12	*Dendroctonus valens* LeConte	Red Turpentine Beetle	强大小蠹	Coleptera; Scolytidae	America areas	1998, Yangcheng, Qinshui of Shanxi	Shanxi, Shannxi, Hebei, and Henan, etc.
13	*Hyphantria cunea* (Drury)	Fall Webworm, American White Moth	美国白蛾	Arctiidae	North America	1979, Liaoning	Liaoning, Hebei, Shandong, Beijing, Tianjin, Shaanxi, Henan, Jilin, etc.
14	*Achating fulica* (Ferussac)	Giant African Snail	非洲大蜗牛	Achatinidae	The east coast of Africa	1920's-1930's, Xiamen, Fujian Province	Guangdong, Hong Kong, Hainan, Guangxi, Yunnan, Fujian, Taiwan.
15	*Pomacea canaliculata* Spix	Apple Snail, Golden Apple Snail,	福寿螺	Ampull4riidae	the Amazon River basin	1981, Guangdong	Guangdong, Guangxi, Yunnan, Fujian, Zhejiang

Table 3. (Continued)

No.	Scientific name	English name	Chinese name	Category	Origin	Invasion history	Current distribution
		Amazonian Snail					
16	*Rana catesbeiana* Shaw	Bull Frog, American Bullfrog	牛蛙	Ranidae	Eastern North America	1951, Taiwan 1959, Guangdong	Almost all over the southern areas up to Beijing, except Tibet and Hainan

Table 4. Second batch of alien invasive species list in China (2010)

No.	Scientific name	English name	Chinese name	Category	Origin	Invasion history	Current distribution
1	*Lantana camara* L.	Common lantana	马缨丹	Verbenaceae	Tropical America	Late Ming Dynasty, Taiwan	Taiwan, Fujian, Guangdong, Hainan, Hong Kong, Guangxi, Yunnan, Sichuan
2	*Ambrosia trifida* L.	Giant ragweed	三裂叶豚草	Compositae	North America	1930's, Liaoning	Jilin, Liaoning, Hebei, Beijing, Tianjin, etc.
3	*Pistia stratiotes* L.	Water lettuce	大薸	Araceae	Brazil	Late Ming Dynasty, the South of China	Almost all over the southern areas up to the Yellow River

No.	Scientific name	English name	Chinese name	Category	Origin	Invasion history	Current distribution
4	*Solidago Canadensis* L.	Canadian goldenrod	加拿大一枝黄花	Compositae	North America	1935, East China	Zhejiang, Shanghai, Anhui, Hubei, Chenzhou of Hunan, Jiangsu, Jiangxi, etc.
5	*Cenchrus echinatus* L.	Bear grass	蒺藜草	Gramineae	Tropical and subtropical America	1934, Taiwan	Fujian, Taiwan, Guangdong, Hong Kong, Guangxi, South Yunnan, etc.
6	*Parthenium hysterophorus* L.	Common parthenium	银胶菊	Compositae	Texas, U.S. and northern Mexico	1926, Yunnan	Yunnan, Guizhou, Guangxi, Guangdong, Hainan, Hong Kong, Fujian, etc.
7	*Flaveria bidentis* (L.) Kuntze	Coastal plain yellowtops	黄顶菊	Compositae	South America	2000, Tianjin	Tianjin, Hebei, etc.
8	*Chenopodium ambrosioides* L.	Mexican tea herb	土荆芥	Chenopodiaceae	Central and South America	1864, Taiwan	Beijing, Shandong, Shaanxi, Shanghai, Zhejiang, Jiangxi, Fujian, Taiwan, Guangdong, Hainan, Hong Kong, Guangxi, Hunan, Hubei, Chongqing, Guizhou, Yunnan, etc.

Table 4. (Continued)

No.	Scientific name	English name	Chinese name	Category	Origin	Invasion history	Current distribution
9	*Amaranthus spinosus* L.	Thorny amaranth	刺苋	Amaranthaceae	Tropical America	1830s, Macao; 1857, Hong Kong	Shaanxi, Hebei, Beijing, Shandong, Henan, Anhui, Jiangsu, Zhejiang, Jiangxi, Hunan, Hubei, Sichuan, Chongqing, Yunnan, Guizhou, Guangxi, Guangdong, Hainan, Hong Kong, Fujian, Taiwan, etc.
10	*Anredera cordifolia* (Tenore) Steenis	Madeira vine, Mignonette vine, Bridal wreath	落葵薯	Basellaceae	South America	1970's, South China or Southwest	Chongqing, Sichuan, Guizhou, Hunan, Guangxi, Guangdong,
						China 1976, Taiwan	Yunnan, Hong Kong, Fujian, etc.
11	*Leptocybe invasa* Fisher et La Salle	Blue gum chalcid	桉树枝瘿姬小蜂	Hymenoptera, Eulophidae	Australia	2007, Guangxi	Guangxi, Hainan, Guangdong

No.	Scientific name	English name	Chinese name	Category	Origin	Invasion history	Current distribution
12	*Lissorhoptrus oryzophilus* Kuschel	Rice water weevil	稻水象甲	Coleoptera, Curculionidae	North America	1988, Hebei	Heibei, Liaoning, Jilin, Shandong, Shanxi, Shaanxi, Zhejiang, Anhui, Fujian, Hunan, Yunnan, Taiwan, etc.
13	*Solenopsis invicta* Buren	Red imported fire ant	红火蚁	Hymenoptera, Formicidae	South America	2003, Taiwan; 2004, Guangdong	Guangdong, Guangxi, Fujian, Hunan, etc.
14	*Procambarus clarkii*	Red swamp crayfish	克氏原螯虾	Decapoda, Cambaridae	North America	1930's, Jiangsu	Xinjiang, Gansu, Ningxia, Inner Mongolia, Shanxi, Shaanxi, Henan, Hebei, Tianjin,
							Beijing, Liaoning, Shandong, Jiangsu, Shanghai, Anhui, Zhejiang, Jiangxi, Hubei, Hunan, Chongqing, Sichuan, Guizhou, Yunnan, Guangdong, Guangxi, Fujian, Taiwan.

Table 4. (Continued)

No.	Scientific name	English name	Chinese name	Category	Origin	Invasion history	Current distribution
15	*Cydia pomonella* (L.)	Codling moth	苹果蠹蛾	Lepidoptera, Tortricidae	Southern Eurasia	1950's, Xinjiang	Xinjiang, Gansu, Inner Mongolia, Heilongjiang, etc.
16	*Liriomyza trifolii* (Burgess)	American serpentine leaf miner	三叶草斑潜蝇	Diptera, Agromyzidae	North America	2005, Guangdong	Taiwan, Guangdong, Hainan, Yunnan, Zhejiang, Yunnan, Jiangsu, Shanghai, Fujian, etc.
17	*Bursaphelenchus xylophilus* (Steiner et Buhrer) Nickle	Pine wood nematode	松材线虫	Aphelenchida, Aphelenchoidi-dae	North America	1982, Nanjing	Jiangsu, Zhejiang, Anhui, Fujian, Jiangxi, Shandong, Hubei, Hunan,
							Guangdong, Chongqing, Guizhou, Yunnan
18	*Hemiberlesia pitysophila* Takagi	Pine armored scale	松突圆蚧	Homoptera, Diaspididae	Japan, Taiwan	1970's, Guangdong	Hong Kong, Macao, Guangdong, Guangxi, Fujian, Jiangxi, etc.

Table 5. Third batch of alien invasive species list in China (2014)

No.	Scientific name	English name	Chinese name	Category	Origin	Invasion history	Current distribution
1	*Amaranthus retroflexus* L.	Redroot amaranth	反枝苋	Amaranthaceae	America	1850's, Hebei, Shandong	Anhui, Beijing, Gansu, Guangdong, Guangxi, Guizhou, Hebei, Henan, Heilongjiang, Hubei, Hunan, Jilin, Jiangsu, Jiangxi, Liaoning, Inner Mongolia, Ningxia, Qinghai, Shandong, Shanxi, Shaanxi,
							Shanghai, Sichuan, Taiwan, Tianjin, Tibet, Xinjiang, Yunnan, Zhejiang, Chongqing.

Table 5. (Continued)

No.	Scientific name	English name	Chinese name	Category	Origin	Invasion history	Current distribution
2	*Aster subulatus* Michx.	Annual saltmarsh aster, bushy starwort	钻形紫菀	Compositae/ Asteraceae	North America	1827, Macao	Anhui, Macao, Beijing, Fujian, Guangdong, Guangxi, Guizhou, Hebei, Henan, Hubei, Hunan, Jiangsu, Jiangxi, Liaoning, Qinghai, Shandong, Shanghai, Sichuan,, Tianjin, Taiwan, Hong Kong, Yunnan, Zhejiang, Chongqing.
3	*Bidens pilosa* L.	Railway beggaricks	三叶鬼针草	Compositae/ Asteraceae	Tropical America	1857, Hong Kong	Anhui, Macao, Beijing, Fujian, Guangdong, Guangxi, Guizhou, Hainan, Hebei, Henan, Hubei, Hunan, Jiangsu, Jiangxi, Shandong, Shanxi, Sichuan,

No.	Scientific name	English name	Chinese name	Category	Origin	Invasion history	Current distribution
							Taiwan, Tianjin, Tibet, Hong Kong, Yunnan, Zhejiang, Chongqing.
4	*Conyza canadersis* (L.) Cronquist	Canadian fleabane, horse-weed	小蓬草	Compositae/ Asteraceae	North America	1860, Shandong	Anhui, Macao, Beijing, Fujian, Gansu, Guangdong, Guangxi, Guizhou, Hainan, Hebei, Henan, Heilongjiang, Hubei, Hunan, Jilin, Jiangsu, Jiangxi, Liaoning, Inner Mongolia, Ningxia, Qinghai, Shandong, Shanxi, Shaanxi, Sichuan,, Tianjin, Taiwan, Tibet, Xinjiang, Hong Kong, Yunnan, Zhejiang, Chongqing.

Table 5. (Continued)

No.	Scientific name	English name	Chinese name	Category	Origin	Invasion history	Current distribution
5	*Conyza bonariensis var. leiotheca* (S.F. Blake) Cuatrec.	Sumatra fleabane, white horseweed	苏门白酒草	Compositae/ Asteraceae	South America	1850's, the South of China	Hubei, Hunan, Jiangsu, Zhejiang, Jiangxi, Fujian, Taiwan, Guangdong, Guangxi,
							Hainan, Hong Kong, Macao, Sichuan, Chongqing, Guizhou, Yunnan, Tibet.
6	*Erigeron annuus* Pers.	Daisy fleabane	一年蓬	Compositae/ Asteraceae	North America	1827, Macao	Almost all over the country except Inner Mongolia, Ningxia, Hainan.
7	*Praxelis clematidea* (Grisebach.) King et Robinson	Praxelis	假臭草	Compositae/ Asteraceae	South America	1980's, Hong Kong	Macao, Fujian, Guangdong, Guangxi, Hainan, Taiwan, Hong Kong, Yunnan.

No.	Scientific name	English name	Chinese name	Category	Origin	Invasion history	Current distribution
8	*Xanthium spinosum* L.	Spiny cocklebur	刺苍耳	Compositae/ Asteraceae	South America	1974, Beijing	Anhui, Beijing, Hebei, Henan, Liaoning, Inner Mongolia.
9	*Ipomoea purpurea* (L.) Roth	Common morning-glory	圆叶牵牛	Convolvulaceae	South America	1890, Mainland China	Anhui, Beijing, Fujian, Gansu, Guangdong, Guangxi, Guizhou, Hainan, Hebei, Henan, Hubei, Hunan, Jilin, Jiangsu, Jiangxi, Liaoning, Inner Mongolia, Ningxia,
							Qinghai, Shandong, Shanxi, Shaanxi, Shanghai, Sichuan,, Tianjin, Taiwan, Tibet, Xinjiang, Hong Kong, Yunnan, Heilongjiang, Zhejiang, Chongqing.

Table 5. (Continued)

No.	Scientific name	English name	Chinese name	Category	Origin	Invasion history	Current distribution
10	*Cenchrus pauciflorus* Benth.	Coast sandbur, common sandbur, field sandbur, spiny bur grass	长刺蒺藜草	Gramineae/ Poaceae	America	1970's, Liaoning, Beijing	Beijing, Shandong, Hebei, Liaoning, Jilin, Inner Mongolia
11	*Trachemys scripta elegans* (Wied.)	Red-eared slider, red eared turtle	巴西龟	Emydidae	The area around the Mississippi River and the Gulf of Mexico, in warm climates in the south-eastern corner of the USA	1980's, Guangdong	Widely distributed in 22 provinces of China with a wild geographic scope of northernmost Liaoning in Shenyang, Southernmost Wuzhishan in Hainan and westernmost Gaoligongshan in Yunnan

No.	Scientific name	English name	Chinese name	Category	Origin	Invasion history	Current distribution
12	*Pterygoplichthys pardalis* (Castelnau)	Amazon sailfin catfish	豹纹脂身鲶	Loricariidae	South America	1980, ornamental fish market	Guangdong, Hubei, Taiwan, Guangxi, Shaanxi, Sichuan, Chongqing, Jiangsu, Jiangxi, Hainan, Anhui, Shanghai, Zhejiang, Fujian, Yunnan, Jilin.
13	*Pygocentrus nattereri* Kner 1858	Red piranha	纳氏臀点脂鲤	Serrasalmidae	The Amazon River Basin of South America	1980's, ornamental fish market	Guangdong, Guangxi, Zhejiang, Sichuan, Hunan, Jiangxi, Beijing, Tianjin, Liaoning, Jilin, Fujian, Hainan, Taiwan.
14	*Oreochromis niloticus* (L.)	Nile tilapia	尼罗罗非鱼	Cichlidae	the Nile Valley	1978, Nanjing	Guangdong, Guangxi, Hainan, Fujian, Taiwan, etc.
15	*Rhynchophorus ferrugineus* (Oliver)	Red palm weevil	红棕象甲	Curculionidae	South Asia, Melanesian islands	1998, Hainan	Hainan, Guangdong, Guangxi, Taiwan, Yunnan, Tibet,

Table 5. (Continued)

No.	Scientific name	English name	Chinese name	Category	Origin	Invasion history	Current distribution
							Jiangxi, Shanghai, Fujian, Sichuan, Guizhou, Jiangsu, Zhejiang.
16	*Corythucha ciliata* Say	Sycamore lace bug	悬铃木方翅网蝽	Tingidae	The east-central of North America	2006, Hubei	Shanghai, Zhejiang, Jiangsu, Chongqing, Sichuan, Hubei, Guiyang, Henan, Shandong, etc.
17	*Phenacoccus solenopsis* Tinsley	Solenopsis mealy bug	扶桑绵粉蚧	Pseudococcidae	North America	2008, Guangdong	Zhejiang, Fujian, Jiangxi, Hunan, Guangdong, Guangxi, Hainan, Sichuan, Yunnan, Hainan, etc.
18	*Quadrastichus erythrinae* Kim	Erythrina gall wasp	刺桐姬小蜂	eulophidae	Mauritius, Reunion, Singapore, USA (Hawaii)	2005, Shenzhen	Taiwan, Guangdong, Guangxi, Fujian, Hainan.

Table 6. The main alien invasive and potential nuisance aquatic animals in China

No.	Scientific name	English name	Chinese name	Category	Origin	Invasion history	Current distribution
1	*Pomacea canaliculata* Spix	Apple snail, Golden apple snail, Amazonian snail	福寿螺	Ampullariidae	the Amazon River basin	1981, Guangdong	Guangdong, Guangxi, Yunnan, Fujian, Zhejiang
2	*Rana catesbeiana* Shaw	Bull frog, American bullfrog	牛蛙	Ranidae	Eastern North America	1951,Taiwan 1959, Guangdong	Almost all over the southern areas up to Beijing, except Tibet and Hainan
3	*Procambarus clarkii*	Red swamp crayfish	克氏原螯虾	Cambaridae	North America	1929, Nanjing	Xinjiang, Gansu, Ningxia, Inner Mongolia, Shanxi, Shaanxi, Henan, Hebei, Tianjin, Beijing, Liaoning, Shandong, Jiangsu, Shanghai, Anhui, Zhejiang, Jiangxi, Hubei, Hunan, Chongqing, Sichuan, Guizhou, Yunnan, Guangdong, Guangxi, Fujian, Taiwan.
4	*Trachemyss cripta elegars* (Wied.)	Red-eared slider, Red eared turtle	巴西龟	Emydidae	The area around the Mississippi River and the Gulf of Mexico, the	1980's, Guangdong	Widely distributed in 22 provinces of China with a wild geographic scope of northernmost Liaoning in Shenyang, Southernmost

Table 6. (Continued)

No.	Scientific name	English name	Chinese name	Category	Origin	Invasion history	Current distribution
					South-eastern corner of the USA		Wuzhishan in Hainan and westernmost Gaoligongshan in Yunnan.
5	*Pterygoplichthys pardalis* (Castelnau)	Amazon sailfin catfish	豹纹脂身鲶	Loricariidae	South America	1980, Guangdong	Guangdong, Hubei, Taiwan, Guangxi, Shaanxi, Sichuan, Chongqing, Jiangsu, Jiangxi, Hainan, Anhui, Shanghai, Zhejiang, Fujian, Yunnan, Jilin.
6	*Pygocentrus nattereri* Kner 1858	Red piranha	纳氏臀点脂鲤	Serrasalmidae	The Amazon River	1980's, Guangdong	Guangdong, Guangxi, Zhejiang, Sichuan, Hunan,
					Basin of South America		Jiangxi, Beijing, Tianjin, Liaoning, Jilin, Fujian, Hainan, Taiwan.
7	*Oreochromis niloticus* (L.)	Nile tilapia	尼罗罗非鱼	Cichlidae	the Nile Valley	1978, Nanjing	Guangdong, Guangxi, Hainan, Fujian, Taiwan, etc.
8	*Clarias gariepinu s*	North African catfish	革胡子鲶	Clariidae	Egypt	1981, Guangdong	South China, Yangtze River, and waters south of the Yangtze River.
9	*Gambusia affinis*	Mosquitofish	食蚊鱼	Poeciliidae	North and Central America	1924, Hangzhou	South China, Yangtze River, and waters south of the Yangtze River.

No.	Scientific name	English name	Chinese name	Category	Origin	Invasion history	Current distribution
10	*Lepisosteus osseus*	Longnose gar	雀鳝	Lepisosteidae	North and Central America	1980, Guangdong	Guangdong, Shanghai, Nanjing, Chongqing
11	*Oreochromis mossambicus*	Mozambique tilapia	莫桑比克罗非鱼	Cichlidae	Africa	1946, Taiwan 1956, Guangdong	Guangdong, Guangxi, Hainan, Fujian, Taiwan, etc.
12	*Oreochromis aureus*	Blue tilapia	奥利亚罗非鱼	Cichlidae	Africa and Eurasia	1981, Guangdong	Mainly in Guangdong, Guangxi and some Southern coastal areas of China.
13	*Carassius cuvieri*	Japanese white crucian carp	日本白鲫	Cyprinidae	Japan	1959, Taiwan 1976, mainland China	Most of the fresh waters in China
14	*Cirrhinus mrigala*	Mrigal carp	麦穗加拉鲮鱼	Cyprinidae	Southeast Asia	1982, Guangdong	fresh waters in South China
15	*Labeo rohita*	Roho labeo	露斯塔野鲮	Cyprinidae	Myanmar	1978, Guangdong	fresh waters in South China
16	*Piaractus brachypomus*	Pirapitinga	短盖巨脂鲤	Serrasalmidae	South America	1982, Taiwan 1985, Guangdong	Most of the provinces in the South of the Yellow River
17	*Micropterus salmoides*	Largemouth black bass	大口黑鲈	Centrarchidae	North America	1983, Shenzhen, Huiyang and Foshan, etc. in Guangdong	Guangdong, Jiangsu, Zhejiang, Sichuan and Taiwan, etc.
18	*Sciaenops Ocellatus*	Red drum	眼斑拟石首鱼	Sciaenidae	USA	1991, Shandong	Coastal waters in Shandong

Table 6. (Continued)

No.	Scientific name	English name	Chinese name	Category	Origin	Invasion history	Current distribution
19	*Lepomis macrochirus*	Bluegill	蓝鳃太阳鱼	Centrarchidae	North America	1987, Hubei	South of the Yangtze river
20	*Clarias batrachus*	Philippine catfish	蟾胡子鲶	Clariidae	Thailand	1978, Guangdong	Some fresh waters and aquariums in South China
21	*Ictalurus punctatus*	Channel catfish	斑点叉尾鮰	Ictaluridae	North America	1984, Hubei	Cultured in over 20 provinces but mainly in Hunan, Hubei, Jiangxi, Anhui, Sichuan, Jiangsu
22	*Mytilopsis sallei* (Recluz, 1849)	Black-striped mussel	沙筛贝	Dreissenidae	Central America	1960s, Taiwan 1980, Hong Kong	Fujian, Guangdong, Guangxi, Hainan, Hong Kong and Taiwan, etc.
23	*Argopecten irradians*	Atlantic bay scallop	海湾扇贝	Pectinidae	USA	1982,Shandon g	Shandong, Liaoning, Hebei
24	*Haliotis rufescens*	Red abalone	红鲍	Haliotidae	USA	1985, Liaoning	Shandong, Liaoning
25	*Haliotis fulgens*	Green abalone	绿鲍	Haliotidae	USA	1985, Liaoning	Shandong, Liaoning
26	*Cherax quadricarinatus*	Redclaw crayfish	红螯螯虾	Parastacidae	Australia	1990, Guangdong, Jiangxi	Guangdong, Jiangxi
27	*Penaeus stylirostris*	Blue shrimp	南美蓝对虾	Penaeidae	South America	2000, Shandong,	Shandong, Jiangsu, Zhejiang
28	*strongylocentrotu s intermedius*	Sea urchin	虾夷马粪海胆	Strongylocentr -otidae	Japan	1989, Liaoning	Liaoning

No.	Scientific name	English name	Chinese name	Category	Origin	Invasion history	Current distribution
29	*Halocynthia roretzi*	Sea squirt	真海鞘	Pyuridae	Japan, Korea	2006, Liaoning, Shandong	Coast of Liaoning and Shandong
30	*Macrochelys temminckii*	Alligator snapping turtle	大鳄龟	Chelydridae	USA	1999, Guangzhou	Beijing, Shanghai, Jiangsu, Zhejiang, Jiangxi, Hainan, Guangdong, Hunan, Shandong, Sichuan, etc.
31	*Chelydra serpentina*	Snapping turtle, North American snapping turtle	拟鳄龟	Chelydridae	USA	1996, Guangzhou	Beijing, Shanghai, Jiangsu, Zhejiang, Jiangxi, Hainan, Guangdong, Hunan, Shandong, Sichuan, etc.

All of the 52 species, over one fifth of which are aquatic organisms, composed of 8 animals and 3 plants, are devastating invasive aliens in the past ten years and the number is still growing.

The greatly increased international trade in the last 30 years has triggered enormous economic growth, as well as environmental threats from an expanding list of non-native species. 529 alien invasive species were put on the government Ministry of Agriculture's registries (Ju et al. 2012) and 544 exotic nuisance species were confirmed in *the second International Conference on Biological Invasions* in 2013. In agricultural and forestry production, 13 main invasive alien species have caused notable direct economic losses of 50 billion RMB every year, according to one estimate. In the last ten years, over 20 dangerous and explosive species, many of them are aquatic organisms including water hyacinths, golden apple snail, bull frog, red-eared slider, red piranha, etc., have successively invaded China with an annual increase of 1-2 species (Su 2013). A nationwide investigation reported that there were 38 aquatic alien invasive animals in 2003 (Xu et al. 2006), but it increased to 52 in 2008 (Xu & Qiang 2011). Water hyacinth (*Eichhornia crassipes*) and golden apple snail (*Pomacea canaliculata*), which are currently distributed in 19 provinces and 15 provinces, respectively, in Mainland China, are two of the most disruptive non-indigenous aquatic species. They still have a significantly negative effect on both economy and ecology, though billions have been spent on their emergency treatment every year.

According to their impacts on agriculture production, biodiversity, human health, and risk cost, we list 31 aquatic animal species as the most invasive or potentially invasive aquatic aliens (Table 6). Obviously, we face an uphill battle of aquatic invasion management and biodiversity conservation in aquatic ecosystems, due to a wide invader distribution in category, seldom unpunished invasion resulting actions, and water as an easy diffused medium.

In ecology, alien invasive species is not only a geological definition, but also more of an ecosystem-based concept. Yunnan province, with 432 fish species, is the most abundant of freshwater fish resources in China. However, this advantage has been totally changed, mainly by biological invasion, which came from two introductions of non-native fish species of Asian carps, stone moroko (*Pseudorasbora parva*), tidewater goby (*Eucyclogobius newberryi*), etc. in 1963-1970 and *Neosalanx taihuensi*, Asian pencil halfbeak (*Hyporhamphus intermedius*), etc., in 1982-1983 respectively. Consequently, over 130 fish species, which is 30 percent of the total, were not collected because of reduction or extinction in the prior five years (Chen et al. 2009). By contrast, up to 25 exotic fish species were recorded during the survey of 11 lakes in 2007-2008 (Yuan et al. 2010). The investigation of fish resources in Lake Fuxian, in the Yunnan province, showed that 11 indigenous species were not collected in 2005 compared to 1995 (Fei et al. 2006).

The extinction of Yunnan lake newt (*Cynops wolterstorffi*) is closely attributed to the introduction of exotic aquatic organisms. Changes of the relative biomass are a reflection of intense interspecific competition between the native and non-native fishes. *Neosalanx taihuensi* was brought to Lake Fuxian from Lake Xingyun by flood in 1982, but it yielded up to tens of tons less than that of the indigenous *Zacco taliensis* 350-400 tons in 1986, then the invader grew to 600-800 tons, compared to 250 tons of the native *Zacco taliensis* in 1990. An investigation in 2005 found that the community structure and spatial distribution of crustacean zooplankton have been greatly changed (Pan et al. 2009). Nowadays, it is hard to catch *Zacco taliensis* due to 10 years of over-fishing, as an additional disturbance. The monitoring strategy should be constructed based on the distribution and damage of the species with additional attention to the large range influence on the aquatic systems.

MANAGEMENT OF ALIEN INVASIVE AQUATIC ANIMALS

Although the biological invasion research is devoted to managing or reducing the amount of aquatic animals present in a body of water, we are also aware that some of the aquatic species have many important functions in aquaculture and in aquatic systems. In China, most of the invasive aquatic animals that harmed the aquatic systems are important aquatic food and some of them have a positive effect as components of the environment. Additionally, the masses can have difficulty sensing the adverse effects caused by direct aquatic invasion. On the other hand, problems caused by excessive aquatic invasion are obvious, and it will be a disaster if we ignore the fact. Therefore, management is of great importance to aquatic invasion, and the management of a body of water should sometimes be to achieve a balance instead of eliminating the invaders (Lopian 2005). For aquatic alien species, prevention is a critically key point, and constructing a comprehensive technical management system of prevention and control may consist of preventative control methods, mechanical control methods, biological control methods, habitat alteration methods, and chemical control methods (Finnoff et al. 2007; Lembi 2000).

Prevention and Control of Invasive Fish

Fish invasion has created serious economic and ecological damage, and the ichthyologists and managers are concerned. However, current understanding of fish invasion and the development of management strategy lag well behind comparable knowledge for other taxonomic groups such as plants and birds (Veltman et al. 1996; Goodwin et al. 1999). Most introductions of fish are deliberate and often related to the economic, sporting, and aesthetic benefits. After the introduction, establishment, and spread, invasive fishes may lower native biodiversity (Witte et al. 2000), alter the structure of the food web (Simon & Townsend 2003), and degrade habitat quality through sediment re-suspension and eutrophication (Zambrano et al. 2001).

Some research has been taken to assess the invasive risk of non-indigenous fish. For example, Copp et al. (2005) developed a risk assessment protocol for the invasive fishes. In general, risk assessment should include two levels: firstly, to conduct the assessment for species not already present in a country and region, and secondly, to address the risk posed by further transfer and spread of already introduced species. Simberloff (2005) argued that the management of exotic species should focus on presumption of risk rather than risk assessment.

Control measures

Physical methods: Complete eradication of fish by physical removal is generally considered impossible given the huge effective effort needed. However, according to the theory of the Allee effect, holding the population at a sufficiently low density for stochastic extinction rather than removing every individual may be the effective method (Courchamp et al. 1999). Physical removal generally intends to lower the ecological impact of invasive fish rather than complete eradication, but it is probably the only effective method in large water bodies because of the impracticality or cost of chemical methods. Common methods are capture techniques such as netting, angling, or electrofishing.

Freshwater fish generally have poor dispersal ability. Waterfalls, chutes and riffles, and the saline waters of estuaries and coasts all possibly prevent the spread of fish. Although constructed physical barriers, like dams, are likely to be a problem for fish passage, it may help to prevent the spread of invasive fishes. Research should be undertaken to prevent the spread of invasive fish while allowing the upstream or downstream passage of migratory indigenous species.

Chemical control: Chemical control is the most common and effective method in eradicating and controlling exotic nuisance fish species (Nicholas 2009). However, due to the potential negative to the ecosystem and environment, this method is limited to small rivers and streams. Four pesticides, namely rotenone, antimycin, saponins, and 3-trifluoromethyl-4-nitrophenol, have been extensively used to remove pest fish. Rotenone, either in liquid or powdered formulations, has been used for many years primarily for the elimination of pest fish from enclosed water bodies. The USA is the largest user of rotenone for fish management (McClay 2002). Since the 1970s, the use of antimycin has declined significantly because of cost and availability. Antimycin is highly toxic to scaled fish, less toxic to scaleless fish, and relatively non-toxic to other aquatic animals. Saponins are a group of chemicals mainly derived from plants. Teaseed cake and mahua oil cake have been widely used in Asian countries. 3-trifluoromethyl-4-nitrophenol (TFM) includes a liquid concentrate, which is applied by boat or backpack sprayers, and solid block, which in secured in a stream. The major requirement for selecting pesticides is that they have strong specificity towards fish but limited toxicity to other species.

Biological control: Most potential bio-control measures are still in the stage of experiment or theory. Some successful applications of natural animals in lab or experimental environmental conditions are less likely to be applicable to natural conditions. Given the relatively non-specific feeding of most fish, and that a predatory species is unlikely to completely eliminate its prey, the introduction of predatory fish seem unlikely to be effective. Moreover, the introductions of exotic piscivorous fish may result in significant ecological problems when no other indigenous species exist to control these exotics.

Pheromone traps are an area of high development in the control of invasive species. They have demonstrated the potential to enhance the capture of pest fish and can be used as a tool to detect nuisance fish incursions, given the extreme sensitivity of some fish to pheromonal compounds (Li et al. 2007). Pheromone technology has been investigated for round-goby (Murphy et al. 2001), goldfish and common carp (Sisler 2005). Using specific pathogens (bacteria, virus, etc.) to control invasive fish is currently experimental. For example, specific carp herpes virus (CyHV-3) shows some potential to limit common carp (McColl et al. 2007) but this disease is highly temperature-dependent and not universally lethal (Gilad et al. 2003).

Management of Invasive Invertebrate: Apple Snail

Some of the most intractable and serious invaders are invertebrates (e.g. woolly adelgid and gypsy moth, Lovett et al. 2006; earthworms, Holdsworth et al. 2007; ants, Holway et al.2002; and snails, Cowie 2002). They make up 26% of the listed 100 world's worst invaders (17 terrestrial, 9 aquatic; Lowe et al. 2000) and 20% (10% for aquatic and terrestrial invertebrates, respectively) of the 485 species listed on the Global Invasive Species Database (GISD; www.issg.org/database).

The golden apple snail, *Pomacea canaliculata* (Lamarck, 1822) (Gastropoda: Ampullariidae), originates from South America and currently has become a highly destructive and quickly spreading group of aquatic invaders (Joshi & Sebastian 2006). This species is the only freshwater snail listed in the global top 100 invasive species (Lowe et al. 2000) and has invaded many countries of East Asia (the mainland of China, Japan, the Republic of Korea, the Philippines, Indonesia, Malaysia, Singapore, Vietnam, Cambodia, Lao PDR, Thailand and Myanmar) and North America (Joshi & Sebastian 2006). It has caused damage in the millions for Asian rice (Halwart 1994) and contributed to an ecological shift of Asian wetland from aquatic plant-based to an undesirable algae-based condition (Carlsson 2004). In addition, *P. canaliculata* can eutrophy water bodies through its high rates of herbivory, defecation, and subsequent nutrient release (Carlsson & Lacoursiere 2005). Recently, it has also been documented that this species, interplaying with climate change, resulted in the emergence of angiostrongliasis in the People's Republic of China (lv et al. 2009). Therefore, in addition to the agricultural and environmental impacts posed by *P. canaliculata*, public health concern is another reason for considering this species as pest invaders.

Control measures

Considering the ecological, agricultural, and health problems resulted from *P. canaliculata*, prevention of introduction and control of spread is probably the only way to avoid the future damage and costs (Naylor 1996). Quarantine programs may help limit the transport of snails from infested to uninfested areas. Additionally, raising the public awareness is also an important step to prevent further spread of *P. canaliculata*.

In general, it is extremely difficult to eradicate apple snails from areas where they have been widely established. Use of molluscicides over large areas is expensive and probably generates negative effects on human health and the environment. Like the control of invasive fish, biological control using predators usually only limits pest numbers to acceptable levels, and does not eradicate the pest. Furthermore, in contrast to the original perception as an environmentally acceptable alternative to the chemicals, biological control is increasingly seen as ecological dangerous (Simberloff & Stiling 1996). Physical measures such as hand-picking are likely to be a relatively effective and appropriate method to control *P. canaliculata* over small areas. Nevertheless, with the increase in areas a large amount of efforts should be involved. Some measures for controlling *P. canaliculata* in China are listed in Table 7 (Yu et al. 2001).

Chemical measures: The main synthetic chemicals used to control the apple snail include copper sulphate, calcium cyanamide, sodium pentachlorophenolate, niclosamide, various organo-tin compounds, endosulphan, metaldehyde, cartap hydrochloride and isazophos (Halwart 1994). Tin-based chemicals were widely applied in the Philippines, but have been banned there because of serious ecological and environmental harm. Copper sulphate is highly toxic to most invertebrates and has potential ecological consequences when they are applied on a broad scale (Cowie 1994). Endosulphan is highly toxic to fish and has been banned in Japan (Halwart 1994). To kill newly hatched snails a second application, one month after the first, will be necessary given that eggs of apple snails, laid above water, will not be affected and new snails probably have been hatched (Cowie 1994). Some chemicals presently in common use in China are listed in Table 7. Nevertheless, some of them, as much as 70% Bayluscide and 65% Sodium pentachlorophenol, are highly toxic to fish. So, overuse should be banned. In Japan, only two of the registered chemicals, i.e. nitrolime and IBP, are used to kill apple snails.

Some research has been undertaken to assess the effect of above pesticides (Palis et al. 1996; Schnorbach 1995) and the environmental impacts inserted have also been increasingly recognized and some efforts have been undertaken. In addition, human health problems are also considered to be associated with the use of pesticides. Based on these considerations, some 'natural' pesticides were researched as an alternative to chemical pesticides.

Biological control: Various animals such as common carp, ducks, lizards, frogs and ants have been shown to feed on *P. canaliculata* or their eggs (Halwart 1994). In a ditch, release of natural animals to control the apple snail is effective when the use of chemical and physical methods is difficult or impossible. Some animals such as ducks (*Anas platyrhynchos*), terrapin (*Amyda japonica*), and common carp (*Cyprinus carpio*) have already been used to control the apple snails.

Table 7. Strategies and measures for controlling golden apple snail (*Pomacea canaliculata*) adopted by farmers in China

Controlling methods	Periods of controlling	Detailed measures	Aims and effects
Agricultural control	Overwintering period	Clearing silt and aquatic plants	Lowering survival rate
		Collecting and removing adult snails in ditches and paddies	Reducing population density
	Emergence period	Clearing egg masses in spring and autumn	Preventing population establishment
		Eliminating adult snails during ploughing period	Lowering density
		Draining water during hatching period	Increasing mortality rate
		Setting net in the water inlet of paddy	Preventing intruding into paddy
Biological control		Ducking breeding after harvesting rice	Lowering density
Chemical control		70% Bayluscide 1 mg/kg	Mortality rate 96.1%
		98% Padan 11-21 mg/kg	Mortality rate 98.0%
		65% Sodium pentachlorophenol 5 mg/kg	Mortality rate 97.9%
		40% Fentin acetate 50 g/667m^2	Mortality rate 92.07%
		98% Copper sulphate 13 mg/kg	Mortality rate 85.90%
		6% Carbaryl-metaldehyde 500 g/667m^2	Mortality rate 80.31%
		6% Metaldahyde 500 g/667m^2	Mortality rate 80.24%

Meanwhile, some of these may cause serious secondary economic or ecological consequence. Some works have been undertaken to evaluate the role of duck and fish as potential biological controllers. The ducks prefer to eat juvenile snails and significantly reduce the snail numbers (Halwart 1994) although they may also affect the rice seedlings. Some research reported the release of common carps in paddies in Taiwan to control the snails and carp are also being promoted in Vietnam for biological control of the apple snail (Pedini & Shehadeh 1996). Since there are ecological consequences of introducing exotic fish, the introduction should be carefully evaluated in terms of both their positive and negative potential impacts and monitored after introduction.

Agricultural and mechanical control: Handpicking, draining water, and transplanting older seedlings are likely to be the most effective non-chemical method to control snails in the rice paddy. Using these measures, most paddies should be able to avoid the apple snails. In the period of transplanting seedlings, handpicking is an effective way to control them. The susceptibility of rice to apple snails declines with seedling age and thus planting seedlings with at least 4-6 weeks old is a recommended method to lower damage (Halwart 1994; Naylor 1996). Constructing slightly deeper strips or ditches within the paddy is an effective method to congregate snails since the paddy is drained and water remains in these ditches. This method facilitates collection of the snails and pesticides treatment. In addition, the use of bait (sacks or nets filled with lettuce, sweet potato leaves, taro leaves) may attract and congregate snails to facilitate handpicking.

Searching for utilization value: The eggs of apple snails could be used as effective additives for pigmentation in some aquarium fish. This is a promising aspect to extend the usage of apple snails to control them. In addition, apple snails have been a naturally effective approach to control weeds. In fact, many farmers in Japan have used them to limit weeds. Additionally, *P. canaliculata* is widely used as a substitute for fishmeal to supply protein in aquaculture and poultry industries in China.

Suggestions for the Future Management

International cooperation and information exchange

The problem of exotic species invasions is trans-boundary in nature, and strong international cooperation and exchange is needed to combat it. Extensive cooperation was conducted to deal with aquatic invasion among European countries and between European and North America. This provides us a great experience and mode regarding how to carry out relevant cooperative subjects or conferences.

In 1998, the International Association of Theoretical and Applied Limnology (SIL) established the Working Group on Aquatic invasive Species

(WGAIS). The main aim of this WGAIS is to develop an information system on aquatic invasive species on a worldwide basis. Some other international and regional working projects include the Nordic Risk Assessment Project that aims to conduct a preliminary risk assessment of alien species in Nordic coastal waters, an EU Concerted Action Project that is used mainly to compare and harmonize various sampling methods of ballast water and study their effectiveness for sampling, The EU MARTOB project assesses potential treatment options for ballast water.

The nature of invasive biology requires broad cooperation and information exchange. This is particular important in the context of globalization considering the increasing introduction rate of exotic species and the large number of surveys and experimental data. Promoting the effective cooperation of Asian or African countries with European or North American countries will help understand and control invasive species.

Institutional improvement

Management system regarding exotic species should be strengthened, particularly for the developing countries. This mainly includes three aspects, e.g., risk assessment systems, permission, and filing systems of exotic introduction. According to the risk assessment system, conducting imparts risk analysis (including hazard identification, risk assessment, risk management and risk communication) to confirm the pest species, its dispersal risk, and possible threatened areas.

Intentional introduction needs a lot of administrative and organizational work, so we need to establish specific institutions to conduct management and arrangement. Introduction units or departments must show detailed materials for the exotic species, including biological and ecological characteristics, introduction history and habitat characteristics of original area, economic benefit analysis, as well as the relevant information of the introduced area. Then, the management department, basing on the comprehensive assessment, decides whether introduction is necessary and appropriate or if additional security measures are needed. When these exotic species have been introduced, a filling system is necessary.

Reinforcing the basic research

Although biological invasion has become a major global issue, our understanding of it is limited, leaving some basic questions to be addressed, such as how to scientifically predict the success, failure of invasion, the relationship between invasion and evolution, as well as the relationship between biological invasion and global changes. Based on the present condition, we believe that the future research point should focus on the following aspects: (1) Evolutionary ecology of invasive species. Even in the last 20 years progress regarding this area has advanced rapidly, and research that is more creative should be carried out. (2) Genomics of invasive species. This is a new research field, using the method of genomics to screen invasive genotypes, and revealing the molecular mechanism of species invasion and evolution. (3) Systematic impacts of species invasion. It is necessary to evaluate the effect of invasion on ecosystem structure and function across multiple spatial and temporal scales for different ecosystems. In the freshwater ecosystem, more attention should be focused on the impacts on biodiversity, aquaculture and shipping, as well as the underlying causal mechanisms.

Managing invasive species in the face of climate change

Climate change and biological invasions are key processes affecting biodiversity, yet their effects have usually been considered separately. In addition to direct roles in habitat quality, climate change will foster the expansion of invasive species into new ranges and magnify the impacts of invaders already present by altering competitive dominance, increasing predation rates, and increasing the virulence of diseases. Resource and conservation managers should consider the interactive effects of climate change and invasive species, thereby establishing management strategies to conserve species and population (Rahel et al. 2008).

Reinforcing education and publicity to improve awareness

According to the needs and ability of different people (relevant technical staff, mangers and ordinary people), some training with different kinds of levels and forms should be conducted to improve preventive and protective awareness. Strengthening other propaganda work and publishing the invasive species list as soon as possible is also necessary to prevent and control invasion.

CONCLUSION

Because of the rapid development of the dramatically international trade, the possibility of spreading invasive alien species increases greatly with consequent enhanced risks. Over 200 exotic aquatic animal species have spread to China through both aquaculture and aquariums. 52 animal species in various aquatic systems have become invaders bringing serious trouble, both current and future. The nuisance of non-indigenous animals is widely distributed in types of habitat, food, web, and category. Not only the transnational species introduction, but also the trans-regional introduction, should be considered in scientific evaluation. The special aquaculture industry in China leads to a great potential threat to system safety concerns in species management and monitoring. Management has done some work to deal with the aquatic invasion, but more work still must be done to upgrade its relevant laws and management regulations to harmonize them with international agreements and standards. Strict laws and regulations for the management of aquatic alien animals are very necessary, as well as more urgency in enhancing awareness of the risks of invasion. More detailed information regarding investigation, surveillance, risk analysis, and control technology should be disclosed to the public. Particular attention should be given to the research of integrated technology in prevention and control, the implementation of rights and obligations of management, international cooperation, and public participation.

ACKNOWLEDGMENTS

We appreciate associate professor Xuejie Wang from Pearl River Fisheries Research Institute for supplying some information of ornamental fish introduction. We also want to thank Mr. Haitao Zhang from Shanghai Ocean University for some information input in manuscript preparation. This work was supported by the funds for the Natural Science Foundation of China (u1131006, 31300468, 31400487) and Agricultural Biological Resources Protection and Utilization Project (2130108).

REFERENCES

Carlsson, N. O. L. & Lacoursière, J. O. (2005). Herbivory on aquatic vascular plants by the introduced golden apple snail (*Pomacea canaliculata*) in Lao PDR. *Biological Invasions*, *7*, 233–241.

Carlsson, N. O. L., Brönmark, C. & Hansson, A. L. (2004). Invading herbivory, the golden apple snail alters ecosystem functioning in Asian wetlands. *Ecology*, *6*, 1575-1580.

Chen, D., Xiong, F., Wang K. & Chang, Y. (2009). Status of research on Yangtze fish biology and fisheries. *Environmental biology of fishes*, *85*, 337-357 (in Chinese with English abstract).

Chen, G., Lin, X. & Chen, P. (2008) Recent progress in the invasion ecology of mosquito fish (*Gambusia spp.*). *Acta Ecologica Sinica*, *28*, 4476-4485 (in Chinese with English abstract).

Congress, U. S. (1990). Nonindigenous aquatic nuisance prevention and control act of 1990. Public Law, 101-646.

Copp, G. H., Garthwaite, R. & Gozlan, R. E. (2005). Risk identification and assessment of non- native freshwater fishes: concepts and perspectives on protocols for the UK. Science Series Technical Report 129. CEFAS, Lowesoft, UK.

Courchamp, F., Clutton-Brock, T. & Grenfell, B. (1999). Inverse density dependence and the Allee effect. *Trends in Ecology and Evolution*, *14*, 405–410.

Cowie, R. H. (2002). Apple snails (Ampullariidae) as agricultural pests: their biology, impacts, and management. In G.M. Baker (Eds), Molluscs as Crop Pests (145–192). Wallingford, CABI Publishing.

Cowie, R. H. (1994). Freshwater snail surveys in relation to eradication of apple snail with copper sulphate on Kauai. Bishop Museum Technical Report, 5, 26.

Ding, J., Mack, R. N., Lu, P., Ren, M. & Huang, H. (2008). China's booming economy is sparking and accelerating biological invasions. *Bioscience*, *58*, 317-324.

Dudgeon, D., Arthington, A. H., Gessner, M. O., Kawabata, Z. I., Knowler, D. J., Lévêque, C. Naiman, R. J., Prieur Richard, A. H., Soto, D. & Stiassny, M. L. (2006) Freshwater biodiversity: importance, threats, status and conservation challenges. *Biological reviews*, *81*, 163-182.

Elton, C. S. (1958). The ecology of invasions by plants and animals. London, Methuen.

Falk-Petersen, J., Bohn, T. & Sandlund, O. T. (2006). On the numerous concepts in invasion biology. *Biological Invasions*, *8*, 1409-1424.

Finnoff, D., Shogren, J. F., Leung, B. & Lodge, D. (2007) Take a risk: preferring prevention over control of biological invaders. *Ecological Economics*, *62*, 216-222.

García-Berthou, E., Alcaraz, C., Pou-Rovira, Q., Zamora, L., Coenders, G. & Feo, C. (2005). Introduction pathways and establishment rates of invasive aquatic species in Europe. *Canadian Journal of Fisheries and Aquatic Sciences*, *62*, 453-463.

Gilad, O., Yun, S., Adkison, M. A., Way, K., Willits, N. H., Bercovier, H. & Hedrick, R. P. (2003). Molecular comparison of isolates of an emerging fish pathogen, koi herpes virus, and the effects of water temperature on mortality of experimentally infected koi. *Journal of General Virology*, *84*, 2661–2668.

Goodwin, B. J., McAllister, A. J. & Fahrig, L. (1999). Predicting invasiveness of plant species based on biological information. *Conservation Biology*, *13*, 422–426.

Halwart, M. (1994). The golden apple snail *Pomacea canaliculata* in Asian rice farming systems: Present impact and future threat. *International Journal of Pest Management*, *40*, 199-206.

Holdsworth, A. R., Frelich, L. E. & Riech, P. B. (2007). Regional extent of an ecosystem engineer:earthworm invasion in northern hardwood forests. *Ecological Applications*, *17*, 1667–1677.

Holway, D. A., Lach, L. S., Suarez, A., Tsutsui N. & Case, T. J. (2002). The causes and consequences of ant invasions. Annual Review of Ecology, *Evolution and Systematics*, *33*, 181–233.

Hulme, P. E. (2009). Trade, transport, and trouble: managing invasive species pathways in an era of globalization. *Journal of Applied Ecology*, *46*, 10-18.

Joshi, R. C. & Sebastian, L. S. (2006). Global advances in ecology and management of golden apple snails. Nueva Ecija, Philippine Rice Research Institute.

Ju, R. T., Li, H., Shih, C. J. & Li, B. (2012). Progress of biological invasions research in China over the last decade. *Biodiversity Science*, *20*, 581-611 (in Chinese with English abstract).

Lembi, C. A. (2000). Aquatic plant management. Purdue University Extension Publication. WS-21-W.

Levine, J. M. (2000). Species diversity and biological invasions: relating local process to community pattern. *Science*, *288*, 852-854.

Li, W., Twohey, M., Jones, M. & Wagner, M. (2007). Research to guide the use of pheromones to control sea lamprey. *Journal of Great Lakes Research*, (special issue 2), 70–86.

Liu, D., Shi, H. T., Liu, Y. X., Wang, J. C., Gong, S. P., Wang, J. & Shen, L. (2011). Distribution status investigation of red-eared slider (*Trachemys scripta elegans*) in China. *Bulletin of Biology*, *46*, 18-21 (in Chinese with English abstract).

Lopian, R. (2005) International plant protection convention and invasive alien species. Proceedings of the workshop on invasive alien species and the international plant protection convention. Braunschweig, Germany. FAO, 22-26.

Lovett, G. M., Canham, C. D., Arthur, M. A., Weathers, K. C. & Fitzhugh, R. D. (2006). Forest ecosystem responses to exotic pests and pathogens in Eastern North America. *Bioscience*, *56*, 395–405.

Lowe, S., Browne, M., Boudjelas, S. & De Poorter, M. (2000). 100 of the World's Worst Invasive Alien Species: A Selection from the Global Invasive Species Database. Auckland, IUCN/SSC.

Lv, S., Zhang, Y., Liu, H. X., Hu, L., Yang, K., Steinmann, P., Chen Z., Wang L. Y., Utzinger J. & Zhou X. N. (2009). Invasive snails and an emerging infectious disease: results from the first national survey on Angiostrongylus cantonensis in China. PLoS neglected tropical diseases, 3, e368.

Mack, R. N., Simberloff, D., Mark Lonsdale, W., Evans, H., Clout, M. & Bazzaz, F. A. (2000). Biotic invasions: causes, epidemiology, global consequences, and control. *Ecological applications*, *10*, 689-710.

McClay, W. (2002). Rotenone use in North America: an update. *Fisheries*, *27*, 19–20.

McColl, K. A., Sunarto, A., Williams, L. M. & Crane, M. St J. (2007). Koi herpes virus: dreaded pathogen or white knight? *Aquaculture Health International*, *9*, 4–6.

Murphy, C. A., Stacey, N. E. & Corkum, L. D (2001). Putative steroidal pheromones in the round goby, Neogobius melanostomus: olfactory and behavioural responses. *Journal of Chemical Ecology*, *27*, 443–470.

Naylor, R. (1996). Invasions in agriculture: Assessing the cost of the golden apple snail in Asia. *AMBIO*, *25*, 443-448.

Nicholas, L. (2009). Management of invasive fish. In: M. N. Clout, P. A. Williams (Eds.), Invasive species management: a handbook of principles and techniques (185–204). Oxford, Oxford University Press.

Palis, F. V., Macatula, R. F. & Browning, L. (1996). Niclosamide, an effective molluscicide for the golden apple snail (*Pomacea canaliculata* Lamarck) control in Philippine rice production systems. *1996 British Crop Protection Council Symposium Proceedings*, *66*, 213-230.

Pan, J. Z., Xiong, F., Li, W. C. & Li, A. Q. (2009). Community structure and spatial distribution of crustacean zooplankton in Lake Fuxian, Yunnan, China. *Journal of Lake Science*, *21*, 408-414 (in Chinese with English abstract).

Pedini, M. & Shehadeh, Z. H. (1996). Projects and other activities. *FAO Aquaculture Newsletter*, *14*, 24-30.

Perrings, C., Dehnen-Schmutz, K., Touza, J. & Williamson, M. (2005). How to manage biological invasions under globalization. *Trends in ecology and evolution*, *20*, 212-215.

Rahel, F. & Olden, J. (2008). Assessing the effects of climate change on aquatic invasive species. *Conservation Biology*, *22*, 521–533.

Richardson, D. M. & Pysek, P. (2008). Fifty years of invasion ecology - the legacy of Charles Elton. *Diversity and Distributions*, *14*, 161-168.

Sakai, A. K., Allendorf, F. W., Holt, J. S., Lodge, D. M., Molofsky, J., With, K. A., Baughman S., Cabin, R. J., Cohen, J. E., Ellstrand, N. C., McCauley, D. E., O'Neil, P., Parker, I. M., Thompson, J. N. & Weller, S. G. (2001). The population biology of invasive species. *Annual Review of Ecology and Systematics*, *32*, 305-332.

Schnorbach, H. J. (1995). The golden apple snail (*Pomacea canaliculata* Lamarck), an increasing important pest in rice, and methods of control with Bayluscid. *Pflanzenschutz-Nachrichten Bayer*, *48*, 313-346.

Simberloff, D. (2005). The politics for assessing risk for biological invasions: the USA as a case study. *Trends in Ecology and Evolution*, *20*, 216–222.

Simberloff, D. & Stiling, P. (1996). How risky is biological control? *Ecology*, *77*, 1965-1974.

Simon, K. S. & Townsend, C. R. (2003). Impacts of freshwater invaders at different levels of ecological organisation, with emphasis on salmonids and ecosystem consequences. *Freshwater Biology*, *48*, 982–994.

Sisler, S. (2005). Behavioural evidence for aggregation pheromones in goldfish (*Carassius auratus*) and common carp (*Cyprinus carpio*). Unpublished PhD thesis, University of Minnesota, MN

State Environmental Protection Administration of China (2003). First batch of alien invasive species list in China. URL: http://www.mep.gov.cn/gkml/zj/wj/200910/ t20091022_172155.htm (in Chinese).

State Environmental Protection Administration of China (2010). Second batch of alien invasive species list in China. URL: http://www.mep.gov.cn/gkml/hbb/bwj/201001/ t20100126_184831.htm (in Chinese).

State Environmental Protection Administration of China (2014). Third batch of alien invasive species list in China. URL: http://www.mep.gov.cn/gkml/hbb/bgg/201408/ t20140828_288367.htm (in Chinese).

Strayer, D. L. (2010). Alien species in fresh waters: ecological effects, interactions with other stressors, and prospects for the future. *Freshwater Biology*, *55*, 152-174.

Su, W. M. (2013). Confirmation of 544 alien invasive species, China has become one of the world's most seriously bioinvasion suffered countries.

URL: http://www.gov.cn/jrzg/2013-10/23/content_2513456.htm (in Chinese).

Valery, L., Fritz, H., Lefeuvre, J. C. & Simberloff, D. (2008). In search of a real definition of the biological invasion phenomenon itself. *Biological Invasions*, *10*, 1345-1351.

Veltman, C. J., Nee, S. & Crawley, M. J. (1996). Correlates of introduction success in exotic New Zealand birds. *The American Naturalist*, *147*, 542–547.

Witte, F., Maku, B. S., Wanink, J. H., Seehausen, O. & Katunzi, E. F. B. (2000). Recovery of cichlid species in Lake Victoria: An examination of factors leading to differential extinction. *Reviews in Fish Biology and Fisheries*, *10*, 233–241.

Xie, Y., Li, Z. Y., William, P. G. & Li, D. M. (2001). Invasive species in China - an overview. *Biodiversity and Conservation*, *10*, 1317-1341.

Xiong, F., Li, W. C., Pan, J. Z., Li, A. Q. & Xia, T. X. (2006). Status and changes of fish resources in Lake Fuxian, Yunnan Province. *Journal of Lake Science*, *18*, 305-311.

Xu, H. G. & Qiang, S. (2011). China's Invasive Alien Species. Beijing, Science Press. 684 pp (in Chinese).

Xu, H. G., Qiang, S., Han, Z. M., Guo, J. Y., Huang, Z. G., Sun, H. Y., He, S. P., Ding, H., Wu, H. R. & Wan, F. H. (2006). The status and causes of alien species invasion in China. *Biodiversity and Conservation*, *15*, 2893-2904.

Yang, Q. L., Cai, L. H., Gao, Y. H., Su, Z., Chen, C. P., Li, B. Q., Xu, C. Y., Lin, G. M., Zheng, H. D. & Xiang, P. (2009). Study on distributional characteristics of organisms in ballast water of exotic ships entering major ports of Fujian and its potential invasion threat. *Transactions of Oceanology and Limnology*, 25-38 (in Chinese with English abstract).

Yu, X. P., Takashi, W., Li, Z. F., Lu, Z. X., Sun, L. P., Zhu, Y. H., Chen, J. M., Zhen, X. S. & Xu, H. X. (2001). Occurrence of golden apple snail, *Pomacea canaliculata* (Lamarck), in paddy fields and its management. *Acta Agriculturae Zhejiangensis*, *13*, 247-252.

Yuan, G., Ru, H. J. & Liu, X. Q. (2010). Fish diversity and fishery resources in lakes of Yunnan Plateau during 2007-2008. *Journal of Lake Science*, *22*, 837-841 (in Chinese with English abstract).

Zambrano, L., Scheffer, M. & Martinez-Ramos, M. (2001). Catastrophic response of lakes to benthivorous fish introduction. *Oikos*, *94*, 344–350.

In: Biological Invasions
Editor: Rebecca Waterman

ISBN: 978-1-63482-019-6

Chapter 4

PHENOLOGY OF *PRUNUS MAHALEB*, A FLESHY FRUITED TREE INVADING NATURAL GRASSLANDS IN ARGENTINE PAMPAS

Martín Raúl Amodeo* and Sergio Martín Zalba
GEKKO (Grupo de Estudios en Conservación y Manejo),
Departamento de Biología, Bioquímica y Farmacia,
Universidad Nacional del Sur, Bahía Blanca, Argentina

ABSTRACT

Natural grasslands are one of the ecosystems that have suffered the most intense series of human changes, even though they have not received enough conservation efforts. The proliferation of invasive shrubs and trees over grasslands and savannas around the world has been one of their severest modifications during the last century. The last remnants of South American Pampas are not the exception, with a set of invasive plants of diverse origins thriving over native biodiversity. Among them, those producing animal-dispersed fruits represent a case of particular interest. St. Lucie's cherry (*Prunus mahaleb* L.) is a small fleshy-fruited tree native to Eurasia that invades natural grasslands in Southern Pampas, where we registered twenty animal species contributing to its dispersal.

---

* San Juan 670, (8000) Bahía Blanca, Argentina. Tel: 54-291-4595101 ext. 2420; Fax: 54-291-4595130. martin.amodeo@uns.edu.ar.

In this chapter, we analyze temporal variations in flowering and availability of ripe fruits at the individual and population level. The species phenology in the area matches with reports from its native range, showing short flowering (19-22 days) and fruiting periods (27-34 days). Synchronicity in flowering and ripening of fruits are consistent with generalist strategies in plant-animal interactions, maximizing cross-pollination probabilities and the attraction of several generalist dispersal agents. We postulate that the lack of reproductive phenological overlapping with similar plants may play a key role in its invasion success. Our results suggest that the protection and shading of other trees might permit the fruits to remain in good condition in the canopies for the frugivores during longer periods. In early January, a massive fruit fall occurs and this might change the main dispersal role from birds consuming fruits in the canopies to mammals eating them directly from the floor, being the formers more related to short-distance dispersal and the latter possibly associated to long distance spread. Big trees with high density of fruits beneath the canopy might act as attraction foci to mammals and might have a key role in promoting long-distance dispersal. This suggests that part of the control efforts should be devoted to manage fruit availability in those attraction foci. *P. mahaleb* reproductive biology and its interactions with dispersal agents can explain to a considerable extent its invasive capacity in the study site and offer valuable information for its management. Incorporating these perspectives on mutualisms and invader-ecosystem interactions might improve the ability to predict invasion processes and contribute to better management decisions in other ecosystems invaded by similar fleshy-fruited species.

INTRODUCTION

Natural grasslands are one of the ecosystems that have suffered the most intense series of human changes, even though they have not received enough conservation efforts (Bilenca & Miñarro, 2004; Hannah et al., 1996). The proliferation of invasive trees and shrubs, at the expense of perennial grasses, has been one of the most striking land cover changes in grasslands and savannas worldwide over the past century, driving major consequences for biodiversity and ecosystem functioning (Archer et al., 2001; Browning et al., 2008; Jurena & Archer, 2003; Richardson et al., 2014; Van Auken, 2009). Natural grasslands from South American Pampas have suffered severe alterations and only a small proportion of its original surface is protected by natural reserves (Bertonatti & Corcuera, 2000; Bilenca & Miñarro, 2004). A set of invasive woody plants of diverse origins spread over native biodiversity in the last remnants of natural grasslands from the Pampas eco-region (Zalba & Villamil, 2002). Among them, those producing animal-dispersed fruits represent a case of particular interest, both because of their ability to rapidly reach different habitats, and also for the opportunity of studying the establishment of novel dispersal interactions.

There are various barriers that an introduced plant must overcome to become invasive in a new environment (Richardson & Pyšek, 2006) and many of these species rely on mutualistic interaction with local species to overcome those barriers (Richardson et al., 2000; Richardson & Pyšek, 2006). Among invasive plants, fleshy-fruited plants dispersed by animals seem to more easily overcome the barriers that limit spread in new habitats, enhancing their chances to become invasive (Buckley et al., 2006; Mooney et al., 2005; Renne et al., 2002). For many species, this achievement seems to be related to the flexibility of their interaction networks with generalist animals. Many vertebrate-dispersed plants exhibit diffuse interactions with a range of dispersal agents (Carlo et al., 2007; Herrera, 1998; Herrera, 2002; Howe, 1993), and this may play a key role for facilitating invasive behavior in regions outside their native range (Buckley et al., 2006; Gosper et al., 2005; Mooney et al., 2005). Fruits of cultivated or spontaneous exotic plants are commonly included in the diet of generalist vertebrates and thus they promote dispersal of nonnative as well as native species (Buckley et al., 2006; Dennis et al., 2007; Gosper et al., 2005; Greenberg & Walter, 2010; Richardson et al., 2000).

Early conceptual contributions to the study of frugivory and seed dispersal emphasized dichotomies in frugivory patterns and fruit characteristics. Certain fruit traits, phenology patterns, quantity and quality of dispersal, and specificity of interactions with frugivores would determine whether plants followed generalized or specialized strategies (McKey, 1975).

According to this view, species with lipid-rich fruits should show asynchronous fruiting patterns to sustain a handful of highly specialized species of frugivores, thus providing high-quality dispersal, while those with watery and sugary fruits should produce superabundant and synchronous fruit crops in order to attract multiple generalist frugivores that provide massive, although low-quality, dispersal (Herrera & Jordano, 1981; Howe, 1993). However, the predictions of this paradigm found little support (Howe, 1993) and evidence has demonstrated that most plant–frugivore systems involve plants sharing multiple frugivores, and frugivores visiting multiple plants and feeding on other organisms at different trophic levels as well (Carlo et al., 2007; Herrera, 1998; Herrera, 2002).

Even though, certain characteristics, as small fruits, high fecundity and short and synchronized phenological events might enable the interaction with a wider variety of dispersers (Corlett, 1998; Gautier-Hion et al., 1985), enhancing their chances of becoming invasive in different habitats (Buckley et al., 2006; Gosper et al., 2005; Gosper & Vivian-Smith, 2009).

The main characteristics of plants that influence its interaction with dispersal agents are those related with spatial and temporal pattern of fruit availability and fruit traits as a food resource (Jordano, 2000). Fruits are offered to dispersers following spatial and temporal patterns that affect the potential ensemble of dispersers, their abundance and foraging movements, and the probability of seeds to be dispersed (Jordano, 2000).

Temporal dimension of those events is critic, because it determines the state reached by and organism or population at a certain moment, and its interaction with other components of the ecosystem. Hence phenology is a very important element that shapes plant-disperser interaction.

The factors that regulate phenology at an individual level contribute also with phenological patterns at the population level (Forrest & Miller-Rushing, 2010). Within a population, the synchronization of phenological events can enhance the probabilities of cross-pollination and the survival of descendents, but also it can increment the competence for resources (Visser et al., 2010). On the other hand, the short duration and synchronism in fruiting at population level results in a superabundant offer of fruits for a wide range of generalist dispersers (Howe, 1993).

St. Lucie's cherry (*Prunus mahaleb* L.) is a small fleshy-fruited tree native to Eurasia that invades grasslands in areas of high value for biodiversity conservation in Southern Pampas. Twenty animal species were registered contributing to its dispersal, offering diverse dispersal services that operate at different spatial scales and habitats.

The study of this interaction is a key aspect for understanding *P. mahaleb* invasiveness. The aim of this chapter is to determine the flowering and fruiting onset and duration within invasive populations from Argentinean natural grasslands, and its relationship with local environmental conditions. It is expected that the phenology of this species coincides with a generalist strategy to attract multiple and generalist dispersal agents, showing short periods and a high synchronization between individuals.

METHODOLOGY

Study Species

Prunus mahaleb L. is a highly light-demanding tree with a prostrate to vertical growth habit (0.5 – 10 m height), depending on local conditions (Kollmann & Pflugshaupt, 2001). Within its native range the species can be found growing on shallow and rather poor soils, from the vicinity of streams to dry hillsides at mid-elevation altitudes (1250-1900 m, Herrera & Jordano, 1981; Jordano, 1994). Corymbs of 3-10 small white flowers, with numerous stamens, are produced by mid spring and bees (Hymenoptera: Andrenidae, Apidae) and flies (Diptera: Calliphoridae, Tachinidae, Syrphidae) are its main pollen vectors (Guitián, Sánchez & Guitián, 1992; Jordano, 1993). Production and ripening of fleshy fruits take place during late spring. One-seeded drupes, bright black when ripe, eight millimeters diameter and with a sugary water-rich pulp are held in auxiliary clusters of one to ten fruits (García *et al.*, 2007; Jordano, 1995). At least 38 bird species, four mammals and one lizard have been recorded feeding on *P. mahaleb* fruits in its native range of distribution (Herrera, 1989; Jordano, 1995; Jordano & Schupp, 2000). In our study area we recorded 20 birds, seven mammals and one ant species manipulating *P. mahaleb* fruits.

Study Area

This study was carried out at Ernesto Tornquist Provincial Park (ETPP), which covers an area of approximately 6700 ha of mountain grasslands in southern Argentine Pampas (38° 3.90'S; 61° 58.33'W), being one of the few protected areas of the whole eco-region (Bilenca & Miñarro, 2004). The climate is temperate, with a mean annual temperature of 14°C and wide annual and daily fluctuations. Annual precipitation is around 800 mm, mostly concentrated in spring and early summer (Burgos, 1968). Vegetation is dominated by grass steppes with scattered rocky outcrops and exotic woody plants growing in small plantations, spontaneous forests or as isolated individuals (Zalba & Villamil, 2002). There are a few native species with fleshy fruits as *Cereus spp.*, *Geoffroea decorticans*, *Margyricarpus pinnatus*, which abundance and distribution are restricted and their dispersal interactions have not been studied (Long & Grassini, 1997). Among exotic woody plants, there are some flehsy-fruited shrubs as *Rosa sp.*, *Rubus ulmifolius* and another

Prunus species, which abundance at the moment is still reduced although many of them are known to be invasive (Zalba & Villamil, 2002). *P. mahaleb* individuals are commonly found, scattered or aggregated in different sized populations in valleys and hillsides. Densest populations are associated with streams and canyons where dispersers' activity seems to be higher. Nevertheless, smaller stands can also be found at mid-elevation on the hill-sides.

Sampling Design

During spring and summer of 2010-2011, the population phenology was analyzed by visiting repeatedly five transects associated with different *P. mahaleb* populations at the study site. During flowering period, transects were transited with intervals of 7-10 days, counting and marking every individual with open flowers (more than 10% of its flowers were open). During fruiting season, the transects were transited with intervals of 10-15 days, registering the number of individuals in four categories defined according to the condition of fruits in its canopy: more than 75% of its fruits were unripe, more than 25% of its fruits were ripe, more than 50% of its fruits were dehydrated and trees with no fruits in their canopy. Four populations were located in open habitats dominated by grasslands (populations 1, 2, 3 and 5) and one population (population 4) was located in a shaded habitat, in the understory of a stand dominated by exotic *Pinus canariensis* of approximately 20-25 m height with a density of 0.045 plants/m^2. In the latter, the dense forest canopy makes a shaded and humid environment in the understory.

At the individual level, phenology was studied intensively in a group of marked individuals growing in open habitat during seasons 2009-2010 and 2010-2011. Within a population, eight trees of different age and size were selected and periodically the total number of fruits in marked branches was counted. Branches were located in different quadrants of the crown (between two and six branches per tree according to its crown volume, representing 1.7 to 8.1% of the total crown volume). Fruits were categorized as unripe, ripe and dehydrated fruits. Counts were made with intervals of 15 days initiating at the end of spring, just before fruit ripening, until no more fruits were found in the trees. Fallen fruits were estimated by using fruit traps installed at 1.5-2.5 m height and distributed randomly under the canopy projection. Traps consisted in bags sustained by a 30 cm diameter wire ring (0.070686 m^2 area).

Elliptic horizontal projection of crown (m^2) was calculated using its maximum diameter and its perpendicular. In order to define an effort proportional to the canopy size, the number of branches and traps were determined for each tree using the formula $2 * \log(canopy\ area)$[1] (Ratiarison & Forget 2011). Between two and six traps were located in each tree, covering between 0.3% y 3.3% of crown horizontal projection. With intervals of 15 days all fruits were counted and extracted within the bags. In this manner, the temporal variation in fruit ripening and its presence in branches and in the ground beneath de tree were analyzed along the entire season.

The total number of fruits produced per plant was estimated by extrapolating the count on each branch to the total canopy considering the proportion of total volume it represented. The total number of fallen fruits beneath the canopy of each tree was also estimated extrapolating the count on each trap to the total canopy area. With these data, the proportion of total fruits that was found on the ground beneath the tree was calculated with respect to the total number of fruits produced initially for five of the trees, in which the effort was more than 1% of canopy volume (branches) and of the canopy area projection (traps).

Data Analysis

In order to assess the variation in phenological events at the population and individual level, Generalized Linear Models were fitted to interpolate the observed proportions between dates, in which the expected value of proportions was related with time by a piecewise-defined function (*spline*, Fan & Yao, 2008). At the population level, two states were considered for flowering period (binomial distribution): flowering trees and trees without flowers with respect to the total number of individuals in the population. Three states were considered for fruiting period (multinomial distribution): individuals with ripe fruits, dehydrated fruits and all trees that did not belong to these categories (those with unripe fruits and those without fruits). For the subset of plants whose fruiting was analyzed individually, three different states were considered with respect to the total number of fruits produced initially in each branch: ripe, unripe and dehydrated fruits (multinomial distribution). Daily changes in proportions were interpolated, defining at the population level the total flowering/fruiting period (more than 10% of the trees with

[1] log= natural logarithm.

flowers/ripe fruits) and the flowering/fruiting peak period at the population level (more than 50% of trees with flowers/ripe fruits); and at the individual level: the total fruiting period (more than 10% of fruits in the branches were ripe) and the fruiting peak period (more than 50% of fruits in branches were ripe).

All analysis were made using R software (R-Core-Team, 2012) and packages *splines*, *stats* (R-Core-Team, 2012), *nnet* (Venables & Ripley, 2002), *gridExtra* (Auguie, 2012) and *ggplot2* (Wickham, 2009).

RESULTS

In populations located in open habitats (1, 2 and 3), the flowering period extended during 19-22 days, from mid-September (16-22/09) until early October (7-11/10). For the population located in the understory of other exotic woody plants, the flowering period extended during 35 days, starting earlier (11/09) and extending to mid-October (16/10). Flowering peak (more than 50% of the population showed open flowers) extended during 13-16 days for populations in open habitats and during 21 days for the population in the understory habitat.

According to focal observations on individuals, plants exhibited open flowers during a period of 14-15 days in open habitats and 21-22 days in the understory habitat. Fruiting period in populations located in open habitats extended during 29-38 days, from late November (19-26/11) until late December (25-28/12). In the population located in the understory habitat, fruiting onset occurred later (3/12) and extended for a longer period (46 days). Fruiting peak (more than 50% of the population with ripe fruits) extended during 24-31 days since late November (23-29/11), except for the population in the understory habitat, for which It was extended during 34 days since mid-December (10/12). By the end of December a high proportion of dehydrated fruits were found in the canopies reducing dramatically the total number of fruits by early January. Some dehydrated fruits remained in the plants until late January and by early February no fruiting plants were found (Table 1, Figure 1).

The basal stem diameter of studied trees varied from 9 to 50 cm, showing total number of fruits in the canopy between 5000 y more than 59000 (Table 2).

Table 1. Flowering and fruiting periods for five *Prunus mahaleb* populations studied during 2010-2011 in Ernesto Tornquist Provincial Park. The total and peak period were defined when more than 10% and 50% of the population exhibited the corresponding phenological state, respectively

	Habitat type (Nr. of populations)	Total period (10%)			Peak period (50%)		
		Start	End	Duration (days)	Start	End	Duration (days)
Open flowers	Open grassland (3)	16-22/09	07-11/10	19-22	19-25/09	5-8/10	13-16
	Understory (1)	[illegible]1/09	16/10	35	19/09	10/10	21
Ripe fruits	Open grassland (4)	19-26/11	25-28/12	29 – 38	23-29/11	22-25/12	24-31
	Understory (1)	3/12	21/01	49	10/12	13/01	34
Dehydrated fruits	Open grassland (4)	21-22/12	13-28/01	22-38	24-26/12	9-28/01	14 - 25
	Understory (1)	11/01	3/02	23	-	-	-

Table 2. Description of trees and sampling effort of *Prunus mahaleb* individuals whose phenology was studied in Ernesto Tornquist Provincial Park

Tree	Basal stem diameter (cm)	Nr. of samples	Branches (% canopy volume)	Traps (% canopy area)	Total number of fruits	SE	Canopy area (m^2)	Fruit density beneath the canopy (fr/m^2)	SE
1	12	3	8.1%	1.3%	13752	(5229)	16.3	537.6	(261)
2	50	6	1.8%	0.3%	59035	(15142.4)	151.6	523.4	(151.7)
3	9.5	2	1.7%	1.6%	12780	(4860)	9.[illegible]	976.2	(424.4)
4	20	3	2.5%	0.5%	32973	(9816)	44.3	547	(188.5)
5	18	2	3.5%	1.5%	8928	(720)	9.2	742.7	(205.1)
6	9	2	4.9%	2.9%	5034	(822)	4.9	205.1	(35.4)
7	42	5	2.8%	0.6%	28488	(7413.4)	62.1	721.5	(119.5)
8	9	2	5.9%	3.3%	5026	(194)	4.2	1082.3	(21.2)

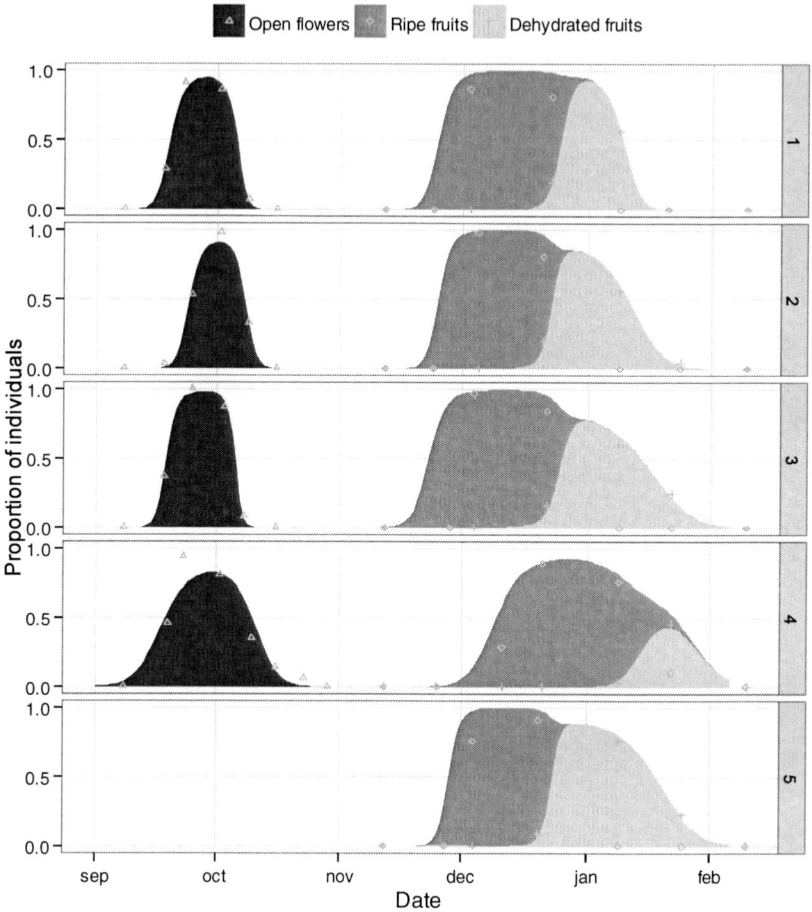

Figure 1. Variation in proportion of individuals with open flowers, ripe fruits and dehydrated fruits along the reproductive season 2010-2011 in five populations of *Prunus mahaleb* located in Ernesto Tornquist Provincial Park. Rhombuses, crosses and triangles indicate observed proportions in each date. Filled areas show modeled proportions.

In open habitats, ripe fruits started to appear in the branches of trees by late November (18/11-02/12), remaining in the canopy during 32-49 days. For all studied trees, fruiting peak (more than 50% of the fruits initially produced were ripe) occurred during 20-32 days within December. The duration of this peak period showed a marginally significant relationship with total canopy area (Simple Linear Regression: $Duration = 20.55 + 2.31 * \log(Canopy\ area)$, $F_{(1;6)}=5.827$, $p=0.0523$, $R^2=0.41$). By early January

(29/12-09/01), the number of fruits in the canopy was dramatically reduced to less than 10% of the total number of fruits produced initially. During the second half of December until first days of January (16/12-06/01), fruit density beneath the trees showed maximum values reaching between values of 42% to 86% of total number of fruits that had fallen during the entire season (mean=55.96%, SE=5.1%, n=8). The presence of fruits beneath the trees extended until late January (22/01–30/01), while around 1% of fruits initially produced remained in the canopy by those days in only five of the studied trees (Figure 2).

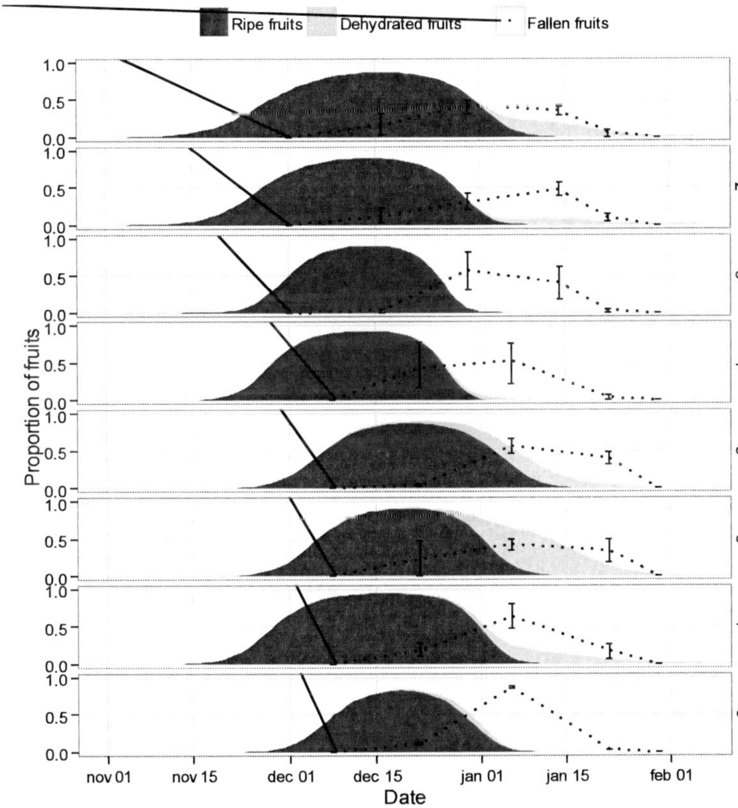

Figure 2. Temporal variation of ripe and dehydrated fruits in the canopy (filled areas) and fruit density beneath the canopy (dotted lines) for eight *Prunus mahaleb* individuals growing in Ernesto Tornquist Provincial Park. Values are expressed as proportions with respect to the total number of fruits initially produced in each tree and with respect to the total number of fallen fruits during the entire season, respectively. Error bars indicate standard error of density values calculated from each trap.

The total number of fallen fruits during the entire season represented between 20% and 91% of the fruits initially produced by each tree (mean= 64.25%, SE=11.9%, n=5). This proportion showed a positive correlation with plant size (logistic regression *%fallen fruits ~ BSD, Likelihood Ratio Test*, Res.Dev.= 6322.2, g.l.=3, $p<0.0001$) but the magnitude of this effect was small (*odds ratio*= 1.07).

Conclusion

The phenology of *Prunus mahaleb* in the study area is characterized by relatively short periods and a high synchronization between individuals and populations. Fruit ripening occurs suddenly and synchronized and, although a small proportion of the fruits remain in some trees during longer periods, the maximum availability of fruits in the canopies occurs during a month by early summer. This situation coincides with similar values reported for other fleshy-fruited plants from temperate ecosystems that ripen during summer-autumn, whose fruiting season extends less than 40 days (0.6-1.3 months, Jordano, 2000). Some of these species show a short fruiting season and their fruits are consumed immediately, like other *Prunus* species, while the fruits of other species persist longer and are consumed along longer periods (Howe & Smallwood, 1982). On the other hand, our results show also a coincidence with the species phenology reported in its native range (Guitián, Sánchez & Rodríguez-Guitián, 1992; Jordano, 1993).

According to the initial paradigm in plant-animal interactions, the short duration and high synchronization of phenological events permit to maximize the probabilities of cross-pollination during flowering and, on the other hand, the attraction of multiple generalist dispersers during fruiting season (Howe, 1993; Jordano, 2000). Mutualisms in the new environment often involve different species that have never coexisted but usually duplicate functions or strategies that exist in the natural range of the plant, emphasizing the diffuse nature of mutualisms (Richardson et al., 2000). These phenological traits, among others, contribute to the invasiveness component in this system, explaining to some extent *P. mahaleb* invasive behavior. Another aspect that arises from our results is the absence of plant species with similar characteristics and abundance in the invaded ecosystem. This represents a key aspect of the invasibility of natural grasslands from the region to fleshy-fruited shrubs as *P. mahaleb*. The introduction of this type of plants represents a qualitative and quantitative change regarding the food resources available to frugivores (Greenberg & Walter, 2010; Lediuk et al., 2014). This situation in which a trait that is new to the resident community of native species offer a competitive advantage to the invasive species has led to the novel weapons hypothesis in biological invasions (Callaway & Ridenour, 2004). This hypothesis has found empirical support mainly for terrestrial plants (Jeschke et al., 2012). We postulate that the lack of reproductive phenological overlapping between P. mahaleb and native species with similar characteristics may play a key role in its invasion success. Overlapping phenology between exotic and native species can result in competition for pollination and dispersing services, reducing the reproductive success (Gosper et al., 2005; Lediuk et al., 2014). On the other hand, when flowering or fruiting phenology of exotic and native species does not overlap, the lack of competition may enhance the seed dispersal favoring the spread of exotic species (Godoy et al., 2009; Gosper, 2004; Gosper et al., 2005). When no other fleshy fruits are available or they are scarce, invasive species can offer a newly available food resource for the local community resulting in more dispersal and faster invasion (Gosper et al., 2005; Greenberg & Walter, 2010; Richardson et al., 2000).

Along the fruiting season, most of the fruits that are produced by a plant fall beneath the canopy. By early January, a large proportion of the fruits in the branches are dehydrated and a massive fruit fall occurs simultaneously in all plants. This phenomenon was also observed during subsequent years and it might be related to daily temperatures, insolation and evapotranspiration increment during this period. Although the dehydrated fruits that persist in the branches are expected to be less attractive for frugivores (Herrera & Jordano, 1981), several feeding events were detected for some bird species during this period, mainly in the inner portion of big trees (personal observation). The fruit offer during this period occurs mainly in the ground beneath the trees. This might change the main dispersal role from birds consuming fruits in the canopies to mammals eating them directly from the floor, being the formers more related to short-distance dispersal and the latter possibly associated to long distance spread (Amodeo & Zalba, 2013; Jordano et al., 2007; Nathan, 2006; Nathan et al., 2008). Short-distance dispersal events contribute to local dynamics of populations and communities and might promote local fruit supply for frugivores. On the other hand, long-distance dispersal events are typically rare, although they play a major role in large-scale processes such as population spread, colonization of new habitats and the flow of individuals between populations (García et al., 2007; Jordano et al., 2007; Nathan, 2006; Nathan, 2007; Nathan et al., 2008). These phenomena have a great conservation concern, such as the spread of invasive fleshy-fruited plants. It has been demonstrated that long-distance dispersal, in spite of being rare and stochastic, are more influential on invasive fleshy-fruited plants spreading processes than short-distance dispersal (Buckley et al., 2006; Kot et al., 1996; Moody & Mack, 1988; Sebert-Cuvillier et al., 2008; Sebert-Cuvillier et al., 2009). Long-distance dispersal theoretical framework states that there exist adaptive traits in plants that favor the occurrence of long distance dispersal events, and hence can be shaped by natural selection (Azihou et al., 2014; Nathan, 2006). Big trees with high density of fruits beneath the canopy might act as attraction foci to mammals and other frugivores that consume the fruits on the ground and might have a key role in promoting long-distance dispersal events. Hence, phenological traits described in this chapter might be favoring the occurrence of long-distance dispersal events mainly during the second half of fruiting season.

The temporal distribution of phenologic events results from a complex interaction between genetic traits of an organism and a set of environmental factors (Forrest & Miller-Rushing, 2010). Large trees showed a fruiting peak slightly longer than small trees. During January, many fruits were observed in the inner part of the canopy while no fruits where observed in the external parts of the canopy (personal observation). The inner portions of the canopy in large trees might offer better protection to insolation and wind, which might permit the fruits to remain in good condition for the frugivores during longer periods. On the other hand, we found some evidence showing that flowering and fruiting seasons tend to be longer in close and shaded habitats. The protection and shading made by other woody plants might explain this phenomenon. It is not uncommon that plants show phenotypic plasticity in their phenology in response to environmental conditions (Visser et al., 2010). The temperature, rainfalls and photoperiod can act as indicators that regulate the inner biological time of an organism or can affect directly the time of biological events (Forrest & Miller-Rushing, 2010). A retardation effect of shading in maturation processes might explain the late appearance of ripe fruits in the close habitat. On the other hand, the offer of ripe fruits in this habitat was longer than in open habitats. These results suggest that the protection of other trees might permit the fruits to remain in good condition in the canopies for the frugivores during longer periods and this facilitation interaction between *P. mahaleb* and other invasive woody plants can be interpreted as a process of invasional meltdown (Simberloff & Von Holle, 1999). The presence of invasive species in an ecosystem facilitates invasion by additional species, increasing their likelihood of survival or ecological impact.

This major hypothesis is considered to play a crucial role in biological invasions and has received good empirical support (Jeschke et al., 2012).

The phenology of *Prunus mahaleb* can explain partially the establishment of dispersal interactions with multiple agents offering a novel food resource that is simultaneously available in different sites during short periods. *P. mahaleb* reproductive biology and its interactions with dispersal agents can explain to a considerable extent its invasive capacity in the study site and offer valuable information for its management. Any management actions that reduce fruit production or affect fruit quality along the fruiting season might affect seed dispersal, both the number of seeds dispersed and the dispersal distribution of those seeds because of differences in frugivores behaviors and their influence on short- and long-distance dispersal events. The strategies for prioritizing management efforts in fleshy-fruited plants have been focused to test whether it is best to remove satellite or core populations, and this might vary with different invasive plant–disperser relationships (Buckley et al., 2006; Gosper et al., 2005). Previous studies have helped us to define that invasive *P. mahaleb* trees should be felled before they reach certain size (basal stem diameter around 9 cm), in order to avoid a dramatic increase in fruit production and local seed dispersal (Amodeo & Zalba, 2013; Zalba & Amodeo, unpublished data). Hence, early detection of isolated trees and rapid eradication of new invasion focuses should be targeted as a principal aim of an effective control strategy. On the other hand, the high accumulation of fruits beneath big trees and its potential effect on long-distance dispersal events discussed here, suggest that efforts should also be devoted to control these attraction foci. Incorporating these perspectives on mutualisms and invader-ecosystem interactions might improve the ability to predict invasion processes and contribute to better management decisions in other ecosystems invaded by similar fleshy-fruited species.

Acknowledgments

This work was funded by CONICET (Consejo Nacional de Investigaciones Científicas y Técnicas, Argentina) and Universidad Nacional del Sur, Argentina.

References

Amodeo, M. & Zalba, S. (2013). Wild cherries invading natural grasslands: unraveling colonization history from population structure and spatial patterns. *Plant Ecology, 214*(11), 1299-1307. doi: 10.1007/s11258-013-0252-4.

Archer, S., Boutton, T.W. & Hibbard., K.A. (2001). Trees in grasslands: biogeochemical consequences of woody plant expansion. In E.D. Schulze, M. Heimann, S. Harrison, E. Holland, J. Lloyd, I.C. Prentice & D.S. Schimel (Eds.), *Global Biogeochemical Cycles in the Climate System* (pp. 115–138). San Diego, California, USA: Elsevier Science.

Auguie, B. (2012). gridExtra: functions in Grid graphics. R package. (Version 0.9.1). Retrieved from http://CRAN.R-project.org/package=gridExtra

Azihou, A., Glèlè Kakaï, R. & Sinsin, B. (2014). Importance of functional traits and regional species pool in predicting long-distance dispersal in savanna ecosystems. *Plant Ecology, 215*(6), 651-660. doi: 10.1007/s11258-014-0330-2

Bertonatti, C. & Corcuera, J. (2000). *Situación ambiental argentina 2000.* Buenos Aires: Fundación Vida Silvestre.

Bilenca, D.N. & Miñarro, F.O. (2004). *Áreas valiosas de pastizal em las pampas y campos de Argentina, Uruguay y sur de Brasil.* Buenos Aires, Argentina: Fundación Vida Silvestre Argentina.

Browning, D.M., Archer, S.R., Asner, G.P., McClaran, M.P. & Wessman, C.A. (2008). Woody plants in grasslands: Post-encroachment stand dynamics. *Ecological Applications, 18*(4), 928-944. doi: 10.1890/07-1559.1

Buckley, Y.M., Anderson, S., Catterall, C.P., Corlett, R.T., Engel, T., Gosper, C.R. et al. (2006). Management of plant invasions mediated by frugivore interactions. *Journal of Applied Ecology, 43*(5), 848-857. doi: 10.1111/j.1365-2664.2006.01210.x

Burgos, J. (1968). El clima de la provincia de Buenos Aires en relación con la vegetación natural y el suelo. In AL Cabrera (Ed.), *Flora de la provincia de Buenos Aires* (Vol. 4, pp. 33-100). Buenos Aires, Argentina: Colección científica INTA.

Callaway, R.M. & Ridenour, W.M. (2004). Novel weapons: invasive success and the evolution of increased competitive ability. *Frontiers in Ecology and the Environment, 2*(8), 436-443.

Carlo, T.s.A., Aukema, J.E. & Morales, J.M. (2007). Plant–Frugivore Interactions as Spatially Explicit Networks: Integrating Frugivore

Foraging with Fruiting Plant Spatial Patterns. In A.J. Dennis, E.W. Schupp, R.J. Green & D.A. Westcott (Eds.), *Seed dispersal: theory and its application in a changing world* (pp. 369-390). UK: CAB International, Columns Design Ltd., Reading.

Corlett, R.T. (1998). Frugivory and seed dispersal by vertebrates in the Oriental (Indomalayan) Region. *Biol Rev Camb Philos Soc, 73*(4), 413-448.

Dennis, A.J., Schupp, E.W., Green, R.J. & Westcott, D.A. (Eds.). (2007). *Seed dispersal: theory and its application in a changing world.* UK: CAB International, Columns Design Ltd., Reading.

Fan, J. & Yao, Q. (2008). *Nonlinear Time Series: Nonparametric and Parametric Methods*. London: Springer.

Forrest, J. & Miller-Rushing, A.J. (2010). Toward a synthetic understanding of the role of phenology in ecology and evolution. *Philosophical Transactions of the Royal Society B: Biological Sciences, 365*, 3101-3112. doi: 10.1098/rstb.2010.0145

García, C., Jordano, P. & Godoy, J.A. (2007). Contemporary pollen and seed dispersal in a *Prunus mahaleb* population: patterns in distance and direction. *Molecular Ecology, 16*(9), 1947-1955.

Gautier-Hion, A., Duplantier, J.M., Quris, R., Feer, F., Sourd, C., Decoux, J.P. et al. (1985). Fruit characters as a basis of fruit choice and seed dispersal in a tropical forest vertebrate community. *Oecologia, 65*(3), 324-337. doi: 10.1007/bf00378906

Godoy, O., Richardson, D.M., Valladares, F. & Castro-Díez, P. (2009). Flowering phenology of invasive alien plant species compared with native species in three Mediterranean-type ecosystems. *Annals of Botany, 103*(3), 485-494. doi: 10.1093/aob/mcn232

Gosper, C.R. (2004). Fruit characteristics of invasive bitou bush, *Chrysanthemoides monilifera* (Asteraceae), and a comparison with co-occurring native plant species. *Australian Journal of Botany, 52*(2), 223-230. doi: http://dx.doi.org/10.1071/BT03046

Gosper, C.R., Stansbury, C.D. & Vivian-Smith, G. (2005). Seed dispersal of fleshy-fruited invasive plants by birds: contributing factors and management options. *Diversity and Distributions, 11*(6), 549 558. doi: 10.1111/j.1366-9516.2005.00195.x

Gosper, C.R. & Vivian-Smith, G. (2009). The role of fruit traits of bird-dispersed plants in invasiveness and weed risk assessment. *Diversity and Distributions, 15*, 1037-1046. doi: 10.1111/j.1472-4642.2009.00599.x

Greenberg, C.H. & Walter, S.T. (2010). Fleshy Fruit Removal and Nutritional Composition of Winter-Fruiting Plants: A Comparison of Non-Native Invasive and Native Species. *Natural Areas Journal, 30*(3), 312-321. doi: 10.3375/043.030.0306

Guitián, J., Sánchez, J.M. & Guitián, P. (1992). Niveles de fructificación en *Crataegus monogyna* Jacq., *Prunus mahaleb* L., *Prunus spinosa* L. (Rosaceae). *Anales Jardín Botánico de Madrid, 50*, 239-245.

Guitián, J., Sánchez, J.M. & Rodríguez-Guitián, M. (1992). Fenología de la floración en un área mediterránea del noroeste ibérico: contrastes con áreas curosiberianas próximas. *Lazaroa, 13*, 111-119

Hannah, L., Carr, J.L. & Lankerani, A. (1996). Human disturbance and natural habitat: a biome level analysis of a global data set. *Biological Conservation, 76*(2), 209-209. doi: 10.1016/0006-3207(96)83209-5

Herrera, C.M. (1989). Frugivory and seed dispersal by carnivorous mammals, and associated fruit characteristics, in undisturbed Mediterranean habitats. *Oikos, 55*, 250-262.

Herrera, C.M. (1998). Long-Term Dynamics Of Mediterranean Frugivorous birds And Fleshy Fruits: A 12-Year Study. *Ecological Monographs, 68*(4), 511-538.

Herrera, C.M. (2002). Seed dispersal by vertebrates. In C.M. Herrera & O. Pellmyr (Eds.), *Plant Animal Interactions: An Evolutionary Approach* (pp. 186-187). Oxford, UK: Wiley.

Herrera, C.M. & Jordano, P. (1981). *Prunus mahaleb* and Birds: The High-Efficiency Seed Dispersal System of a Temperate Fruiting Tree. *Ecological Monographs, 51*(2), 203-218.

Howe, H.F. (1993). Specialized and generalized dispersal systems: where does 'the paradigm' stand? *Vegetatio, 107-108*(1), 3-13. doi: 10.1007/bf00052208

Howe, H.F. & Smallwood, J. (1982). Ecology of Seed Dispersal. *Annual Review of Ecology and Systematics, 13*, 201-228.

Jeschke, J., Gómez Aparicio, L., Haider, S., Heger, T., Lortie, C., Pyšek, P. et al. (2012). Support for major hypotheses in invasion biology is uneven and declining. *NeoBiota, 14*(0), 1-20. doi: 10.3897/neobiota.14.3435

Jordano, P. (1993). Pollination biology of *Prunus mahaleb* L.: deferred consequences of gender variation for fecuntdity and seed size. *Biological Journal of the Linnean Society, 50*, 65-84.

Jordano, P. (1994). Spatial and temporal variation in the avian-frugivore assemblage of *Prunus mahaleb*: patterns and consequences. *Oikos, 71*(3), 479-491.

Jordano, P. (1995). Frugivore-mediated selection on fruit and seed size: birds and St. Lucie's cherry, *Prunus Mahaleb. Ecology, 76*(8), 2627-2639.

Jordano, P. (2000). Fruits and frugivory. In M. Fenner (Ed.), *Seeds: the ecology of regeneration in natural plant communities.* (2nd ed., pp. 125-165). Wallingford, UK: CABI Pub.

Jordano, P., García, C., Godoy, J.A. & García-Castaño, J.L. (2007). Differential contribution of frugivores to complex seed dispersal patterns. *Proceedings of the National Academy of Sciences, 104*(9), 3278-3282. doi: 10.1073/pnas.0606793104

Jordano, P. & Schupp, E.W. (2000). Seed disperser effectiveness: the quantity component and patterns of seed rain for *Prunus mahaleb. Ecological Monographs, 70*(4), 591-615.

Jurena, P.N. & Archer, S. (2003). Woody plant establishment and spatial heterogeneity in grasslands. *Ecology, 84*(4), 907-919. doi: 10.1890/0012-9658(2003)084[0907:wpeash]2.0.co;2

Kollmann, J. & Pflugshaupt, K. (2001). Flower and fruit characteristics in small and isolated populations of a fleshy-fruited shrub. *Plant Biology, 3*(1), 62-71.

Kot, M., Lewis, M.A. & van den Driessche, P. (1996). Dispersal data and the spread of invading organisms. *Ecology, 77*(7), 2027-2042.

Lediuk, K.D., Damascos, M.A., Puntieri, J.G. & Svriz, M. (2014). Differences in phenology and fruit characteristic between invasive and native woody species favor exotic species invasiveness. *Plant Ecology, 215*(12), 1455-1467.

Long, M. & Grassini, C. (1997). *Actualización del conocimiento florístico del Parque Provincial Ernesto Tornquist*. Bahía Blanca: Ministerio de Asuntos Agrarios Provincia de Buenos Aires y Universidad Nacional del Sur.

McKey, D. (1975). The ecology of coevolved seed dispersal systems. In L.E. Gilbert & P.H. Raven (Eds.), *Coevolution of animals and plants* (pp. 159-191). Austin, Texas: University of Texas Press.

Moody, M.E. & Mack, R.N. (1988). Controlling the spread of plant invasions: the importance of nascent foci. *Journal of Applied Ecology, 25*(3), 1009-1021.

Mooney, H.A., Mack, R.N., McNeely, J.A., Neville, L.E., Schei, P.J. & Waage, J.K. (2005). *Invasive alien species: a new synthesis*. Paris, France: Scientific Committee on Problems of the Environment (SCOPE).

Nathan, R. (2006). Long-Distance Dispersal of Plants. *Science, 313*, 786-788. doi: 10.1126/science.1124975

Nathan, R. (2007). Total Dispersal Kernels and the Evaluation of Diversity and Similarity in Complex Dispersal Systems. In A.J. Dennis, E.W. Schupp, R.J. Green & D.A. Westcott (Eds.), *Seed dispersal: theory and its application in a changing world* (pp. 252-276). UK: CAB International, Columns Design Ltd., Reading.

Nathan, R., Schurr, F.M., Spiegel, O., Steinitz, O., Trakhtenbrot, A. & Tsoar, A. (2008). Mechanisms of long-distance seed dispersal. *Trends in Ecology & Evolution, 23*(11), 638-647. doi: http://dx.doi.org/10.1016/j.tree.2008.08.003

R-Core-Team. (2012). R: A language and environment for statistical computing (Version 2.15.1). Vienna, Austria: R Foundation for Statistical Computing. Retrieved from http://www.R-project.org/.

Renne, I.J., Barrow, W.C., Johnson Randall, L.A. & Bridges, W.C. (2002). Generalized avian dispersal syndrome contributes to Chinese tallow tree (*Sapium sebiferum*, Euphorbiaceae) invasiveness. *Diversity and Distributions, 8*(5), 285-295. doi: 10.1046/j.1472-4642.2002.00150.x

Richardson, D., Hui, C., Nuñez, M. & Pauchard, A. (2014). Tree invasions: patterns, processes, challenges and opportunitics. *Biological Invasions, 16*(3), 473-481. doi: 10.1007/s10530-013-0606-9

Richardson, D.M., Allsopp, N., D'Antonio, C.M., Milton, S.J. & Rejmánek, M. (2000). Plant invasions — the role of mutualisms. *Biological Reviews, 75*(1), 65-93. doi: 10.1111/j.1469-185X.1999.tb00041.x

Richardson, D.M. & Pyšek, P. (2006). Plant invasions: merging the concepts of species invasiveness and community invasibility. *Progress in Physical Geography, 30*(3), 409-431.

Sebert-Cuvillier, E., Simon-Goyheneche, V., Paccaut, F., Chabrerie, O., Goubet, O. & Decocq, G. (2008). Spatial spread of an alien tree species in a heterogeneous forest landscape: a spatially realistic simulation model. *Landscape Ecology, 23*(7), 787-801. doi: 10.1007/s10980-008-9237-4

Sebert-Cuvillier, E., Simonet, M., Simon-Goyheneche, V., Paccaut, F., Goubet, O. & Decocq, G. (2009). PRUNUS: a spatially explicit demographic model to study plant invasions in stochastic, heterogeneous environments. *Biological Invasions, 12*(5), 1183-1206. doi: 10.1007/s10530-009-9539-8

Simberloff, D. & Von Holle, B. (1999). Positive Interactions of Nonindigenous Species: Invasional Meltdown? *Biological Invasions, 1*(1), 21-32. doi: 10.1023/a:1010086329619

Van Auken, O.W. (2009). Causes and consequences of woody plant encroachment into western North American grasslands. *Journal of*

Environmental Management, 90(10), 2931-2942. doi: http://dx.doi.org/10.1016/j.jenvman.2009.04.023

Venables, W.N. & Ripley, B.D. (2002). *Modern Applied Statistics with S. Fourth Edition*. New York: Springer.

Visser, M.E., Caro, S.P., van Oers, K., Schaper, S.V. & Helm, B. (2010). Phenology, seasonal timing and circannual rhythms: towards a unified framework. *Philosophical Transactions of the Royal Society B: Biological Sciences, 365*(1555), 3113-3127. doi: 10.1098/rstb.2010.0111

Wickham, H. (2009). ggplot2: elegant graphics for data analysis (Version 0.9.2.1) [package]. New York: Springer.

Zalba, S.M. & Villamil, C.B. (2002). Woody plant invasion in relictual grasslands. *Biological Invasions, 4*(1), 55-72.

INDEX

B

C

D

E

F

G

H

I

J

K

L

M

N

O

P

R

S

T

U

V

W

Z